Electromagnetism and Geometry

Electromagnetism and Geometry

Si Li

Department of Mathematical Sciences

Tsinghua University

International Press of Boston, Inc. · Somerville, Massachusetts, U.S.A.

Electromagnetism and Geometry

by Si Li
Department of Mathematical Sciences, Tsinghua University

Somerville, Massachusetts, U.S.A.
www.intlpress.com

ISBN (print): 978-1-57146-462-0

Printed in China, November 2023

Contents

Preface

In April 2021, Qiuzhen College (求真书院) was newly established at Tsinghua University under the leadership of Professor Shing-Tung Yau. It homes the distinguished elite mathematics program in China starting in 2021: the "Yau Mathematical Sciences Leaders Program" (丘成桐数学科学领军人才培养计划). This program puts strong emphasis on basic sciences related to mathematics in a broad sense. Though majored in mathematics, students in this program are required to study fundamental theoretical physics such as classical mechanics, electromagnetism, quantum mechanics, and statistical mechanics, in order to understand global perspectives of theoretical sciences. It is an exciting challenge both for students and for instructors.

This preliminary note is written for the course "Electrodynamics" that I lectured at Qiuzhen College in the spring semester of 2023. The lecture note consists of two parts. The first part is to explain key physics ingredients of electromagnetism, such as Maxwell's equations, electrostatics, magnetostatics, electromagnetic waves, radiation, scattering, etc. The second part is of geometric nature, which explains Maxwell theory as U(1)-gauge theory in terms of fiber bundle theory, as well as its consistency with special relativity. We emphasize on different faces of concrete examples in order to understand the bridge between physics and mathematics. This note is in succession to [10] in this series, and assumes basic knowledge on differential forms and vector fields. Readers can consult [10] for preliminary geometric backgrounds.

I greatly appreciate the "*Notes Taker Program*" at Qiuzhen College, which has triggered and supported the production of this note. A preliminary version of this note was carefully typed by Yang Peng (杨鹏) all the way along the course, and I am extremely grateful to his great job for *Notes Taker*. I would also like to thank Zhou Jiawei (周嘉伟), Dingxu Zhihan (丁徐祉晗), Liu Jiuhe (刘九和) and Quan Hanwen (权瀚文) for their help on proofreading of the first version of this note.

@ Jingzhai (静斋)
June, 2023

This is a drawing by my daughter expressing herself with her brother. I found it interesting as it illustrates precisely the essence of electromagnetism on the coupling of electric and magnetic fields as well as the topological nature of Maxwell's equations.

Chapter 1

Introduction: Maxwell's Equations

In the early 1860's, James Clerk Maxwell took the work of Faraday and many others, and summarized into four equations that linked the electric field with the magnetic field. Maxwell's equations are nowadays accepted as the basis of all modern theories of electromagnetism.

$$\begin{cases} \nabla \cdot \vec{\mathbf{E}} = \rho/\varepsilon_0 \\ \nabla \cdot \vec{\mathbf{B}} = 0 \\ \nabla \times \vec{\mathbf{E}} = -\frac{\partial \vec{\mathbf{B}}}{\partial t} \\ \nabla \times \vec{\mathbf{B}} = \mu_0 \left(\vec{\mathbf{j}} + \varepsilon_0 \frac{\partial \vec{\mathbf{E}}}{\partial t} \right) \end{cases}$$

These equations predicted electromagnetic waves travelling at the observed speed of light. This leads to Maxwell's speculation that lights are electromagnetic waves, and suddenly brings light, electricity and magnetism into the same fundamental phenomenon.

Maxwell's equations in modern geometry take the concise form

$$\begin{cases} dF = 0 \\ d^*F = J \end{cases}$$

Here F is a 2-form on spacetime that collects both electric and magnetic fields. J is a 1-form that represents the electric charge-current. d is the de Rham differential, and d^* is its adjoint. This immediately reveals many deep geometric and topological natures of Maxwell theory.

In this chapter, we will review and explain the precise meaning of these equations, preparing for the journey toward the study of physics and geometry of electromagnetism. The beauty of this set of equations lies in its the simplicity and nontriviality. For example, it encodes all together the following fundamental electromagnetic phenomenons

stationary charges	$\Longrightarrow$	electric fields
steady moving charges	$\Longrightarrow$	magnetic fields
accelerating charges	$\Longrightarrow$	electromagnetic wave

These will be discussed and explored in detail in Chapter 2 and Chapter 3. Geometric and topological aspects of Maxwell's equations will be explained in Chapter 4 and Chapter 5.

Our flavor will be geometric, and assume basic knowledge on differential forms. Readers can consult [10, Chapter 3] in this series for preliminary geometric backgrounds.

1.1 Hodge Star

Hodge Star on $\mathbb{R}^3$

Let us consider the geometry of $\mathbb{R}^3$ with space coordinate (x, y, z). Denote by

$$d_3 = dx\frac{\partial}{\partial x} + dy\frac{\partial}{\partial y} + dz\frac{\partial}{\partial z}$$

for the de Rham differential on $\mathbb{R}^3$ for space variables. The Euclidean metric $ds^2 = dx^2 + dy^2 + dz^2$ defines a Hodge star operator on differential forms, which we denote by

$$*_3 : \Omega^p(\mathbb{R}^3) \longrightarrow \Omega^{3-p}(\mathbb{R}^3).$$

Here $\Omega^p(\mathbb{R}^3)$ is the space of smooth p-forms on $\mathbb{R}^3$. Explicitly, $*_3$ applies to a basis as

$$\begin{aligned}
1 &\overset{*_3}{\longmapsto} dx \wedge dy \wedge dz \\
dx &\overset{*_3}{\longmapsto} dy \wedge dz \\
dy &\overset{*_3}{\longmapsto} dz \wedge dx \\
dz &\overset{*_3}{\longmapsto} dx \wedge dy \\
dx \wedge dy &\overset{*_3}{\longmapsto} dz \\
dy \wedge dz &\overset{*_3}{\longmapsto} dx \\
dz \wedge dx &\overset{*_3}{\longmapsto} dy \\
dx \wedge dy \wedge dz &\overset{*_3}{\longmapsto} 1
\end{aligned}$$

Then $*_3$ is defined on all $\Omega^\bullet(\mathbb{R}^3)$ by $C^\infty(\mathbb{R}^3)$-linear extension over the above basis. For example,

$$*_3(f dx + g dy \wedge dz) = f dy \wedge dz + g dx$$

where $f, g \in C^\infty(\mathbb{R}^3)$ are functions on $\mathbb{R}^3$. The above $*_3$ has the property that

$$(*_3)^2 = 1 \qquad \text{on all} \quad \Omega^\bullet(\mathbb{R}^3).$$

Definition 1.1.1. We define the adjoint operator d_3^* of d_3 in $\mathbb{R}^3$ by

$$d_3^* = (-1)^p *_3 d_3 *_3 : \Omega^p(\mathbb{R}^3) \longrightarrow \Omega^{p-1}(\mathbb{R}^3).$$

Let ∇^2 be the Laplacian operator defined by

$$\nabla^2 f := \left(\partial_x^2 + \partial_y^2 + \partial_z^2\right) f.$$

∇^2 can be extended to $\Omega^p(\mathbb{R}^3)$ component-wise with respect to the basis as above. For example,

$$\nabla^2(f dx + g dy \wedge dz) = (\nabla^2 f)dx + (\nabla^2 g)dy \wedge dz.$$

We will still denote it by

$$\nabla^2 : \Omega^\bullet(\mathbb{R}^3) \to \Omega^\bullet(\mathbb{R}^3).$$

Proposition 1.1.2. *$d_3 d_3^* + d_3^* d_3$ is related to the Laplacian operator ∇^2 by*

$$d_3 d_3^* + d_3^* d_3 = -\nabla^2.$$

Proof: This is a good exercise. □

As an example for illustration, let f be a function on $\mathbb{R}^3$. Since f is a 0-form, $d_3^* f = 0$. Then

$$\begin{aligned}(d_3 d_3^* + d_3^* d_3)f = d_3^* d_3 f &= - *_3 d_3 *_3 (\partial_x f dx + \partial_y f dy + \partial_z f dz) \\ &= - *_3 d_3(\partial_x f dy \wedge dz + \partial_y f dz \wedge dx + \partial_z f dx \wedge dy) \\ &= - *_3 (\partial_x^2 f + \partial_y^2 f + \partial_z^2 f) dx \wedge dy \wedge dz = -(\partial_x^2 + \partial_y^2 + \partial_z^2) f.\end{aligned}$$

Integral in $\mathbb{R}^3$

Differential p-forms can be integrated on p-dim oriented spaces. Besides, we will also use the following notations for various integrals in $\mathbb{R}^3$ in this note:

- Volume integral of a function f on a 3-dim region $V \subset \mathbb{R}^3$:

$$\int_V d^3 r f(\vec{r}), \qquad \vec{r} = (x, y, z)$$

  Here $d^3 r = dx dy dz$ is the standard volume form on $\mathbb{R}^3$.

- Surface integral of a vector $\vec{A}$ on a 2-dim surface $S \subset \mathbb{R}^3$:

$$\int_S d\vec{\sigma} \cdot \vec{A}$$

  Here $d\vec{\sigma}$ is the vector surface element on S, and

$$d\vec{\sigma} \cdot \vec{A} = d\sigma_x A_x + d\sigma_y A_y + d\sigma_z A_z = dy dz A_x + dz dx A_y + dx dy A_z.$$

- Line integral of a vector $\vec{A}$ on a 1-dim curve $C \subset \mathbb{R}^3$:

$$\int_C d\vec{r} \cdot \vec{A}$$

Here $d\vec{r}$ is the vector line element on C, and

$$d\vec{r} \cdot \vec{A} = dx A_x + dy A_y + dz A_z.$$

- We will sometimes write $\int_Y f$ for the integral of a function f over some space $Y \subset \mathbb{R}^3$, without explicitly expressing the measure. The measure is understood as the canonically induced one from $\mathbb{R}^3$. When Y is 3-dim, this is the volume integral; when Y is 2-dim, this is the area integral; when Y is 1-dim, this is the curve integral.

Example 1.1.3. Let S be a surface in $\mathbb{R}^3$ and $\alpha = f_x dx + f_y dy + f_z dz \in \Omega^1(\mathbb{R}^3)$. We associate a vector field $\vec{A} = (f_x, f_y, f_z)$ by collecting components of α. Then the usual surface integral of $\vec{A}$ with respect to the vector surface element $d\vec{\sigma}$ can be expressed via Hodge star as

$$\int_S d\vec{\sigma} \cdot \vec{A} = \int_S *_3 \alpha.$$

Hodge Star on $\mathbb{R}^{3,1}$

Consider now the Minkowski space $\mathbb{R}^{3,1}$ with space-time coordinate (x, y, z, t). We will simply denote by

$$d = dx\frac{\partial}{\partial x} + dy\frac{\partial}{\partial y} + dz\frac{\partial}{\partial z} + dt\frac{\partial}{\partial t}$$

for the de Rham differential on $\mathbb{R}^{3,1}$. The Minkowski metric $ds^2 = c^2 dt^2 - dx^2 - dy^2 - dz^2$ induces a Hodge star operator

$$* : \Omega^p(\mathbb{R}^{3,1}) \longrightarrow \Omega^{4-p}(\mathbb{R}^{3,1}).$$

Here c is the speed of light. Explicitly, we have

$$\begin{aligned}
1 &\overset{*}{\longmapsto} c\,dt \wedge dx \wedge dy \wedge dz \\
c\,dt &\overset{*}{\longmapsto} dx \wedge dy \wedge dz \\
dx &\overset{*}{\longmapsto} c\,dt \wedge dy \wedge dz \\
dy &\overset{*}{\longmapsto} c\,dt \wedge dz \wedge dx \\
dz &\overset{*}{\longmapsto} c\,dt \wedge dx \wedge dy \\
c\,dt \wedge dx &\overset{*}{\longmapsto} -dy \wedge dz \\
c\,dt \wedge dy &\overset{*}{\longmapsto} -dz \wedge dx
\end{aligned}$$

$$
\begin{aligned}
c\,dt \wedge dz &\overset{*}{\longmapsto} -dx \wedge dy\\
dx \wedge dy &\overset{*}{\longmapsto} c\,dt \wedge dz\\
dy \wedge dz &\overset{*}{\longmapsto} c\,dt \wedge dx\\
dz \wedge dx &\overset{*}{\longmapsto} c\,dt \wedge dy\\
dx \wedge dy \wedge dz &\overset{*}{\longmapsto} c\,dt\\
c\,dt \wedge dx \wedge dy &\overset{*}{\longmapsto} dz\\
c\,dt \wedge dy \wedge dz &\overset{*}{\longmapsto} dx\\
c\,dt \wedge dz \wedge dx &\overset{*}{\longmapsto} dy\\
c\,dt \wedge dx \wedge dy \wedge dz &\overset{*}{\longmapsto} -1
\end{aligned}
$$

It is direct to check that

$$*^2 = (-1)^{p+1} \qquad \text{on } \Omega^p(\mathbb{R}^{3,1}).$$

Definition 1.1.4. We define the adjoint operator d^* of d on $\mathbb{R}^{3,1}$ by

$$d^* = *d* : \Omega^p(\mathbb{R}^{3,1}) \longrightarrow \Omega^{p-1}(\mathbb{R}^{3,1}).$$

Remark 1.1.5. The sign in defining d^* via $*$ on $\mathbb{R}^{3,1}$ is different from that on $\mathbb{R}^3$, due to dimension and signature reason.

Proposition 1.1.6. *In the Minkowski spacetime $\mathbb{R}^{3,1}$, we have*

$$dd^* + d^*d = -\Box,$$

where $\Box$ is the d'Alembert operator

$$\Box = \frac{1}{c^2}\frac{\partial^2}{\partial t^2} - \frac{\partial^2}{\partial x^2} - \frac{\partial^2}{\partial y^2} - \frac{\partial^2}{\partial z^2} = \frac{1}{c^2}\frac{\partial^2}{\partial t^2} - \nabla^2.$$

Here $\Box$ is defined on $\Omega^p(\mathbb{R}^3)$ component-wise with respect to the basis as above, similar to ∇^2.

Proof: Exercise. $\Box$

Remark 1.1.7. Here is a useful formula relating geometric operators in $\mathbb{R}^3$ and in $\mathbb{R}^{3,1}$. Let α be a p-form in $\mathbb{R}^{3,1}$ containing only form indices dx, dy, dz, and $|\alpha| = p$ is the form degree of α. Then

$$
\begin{cases}
*\alpha = cdt \wedge *_3\alpha\\
*(\alpha \wedge cdt) = *_3\alpha\\
d^*\alpha = -d_3^*\alpha\\
d^*(\alpha \wedge cdt) = -d_3^*\alpha \wedge cdt - (-1)^{|\alpha|}\frac{1}{c}\partial_t\alpha
\end{cases}
$$

1.2 Maxwell's Equations

The modern form of Maxwell's equations is

$$\begin{cases} \nabla \cdot \vec{\mathbf{E}} = \rho/\varepsilon_0 \\ \nabla \cdot \vec{\mathbf{B}} = 0 \\ \nabla \times \vec{\mathbf{E}} = -\frac{\partial \vec{\mathbf{B}}}{\partial t} \\ \nabla \times \vec{\mathbf{B}} = \mu_0 \left(\vec{\mathbf{j}} + \varepsilon_0 \frac{\partial \vec{\mathbf{E}}}{\partial t} \right) \end{cases}$$

This set of equations completely describes the dynamics and interactions of electromagnetic fields. Here

$\vec{\mathbf{E}} = (\mathbf{E}_x, \mathbf{E}_y, \mathbf{E}_z)$ is the electric field

$\vec{\mathbf{B}} = (\mathbf{B}_x, \mathbf{B}_y, \mathbf{B}_z)$ is the magnetic field

ρ is the electric charge density

$\vec{\mathbf{j}} = (\mathbf{j}_x, \mathbf{j}_y, \mathbf{j}_z)$ is the electric current density

These quantities in general depend on the space (x, y, z) and time t. The other constants in Maxwell's equations are: ε_0 the permittivity of free space and μ_0 the permeability of free space. They are related to the speed of light c by the following equation

$$c = \frac{1}{\sqrt{\varepsilon_0 \mu_0}}.$$

A particle of charge q moving with velocity $\vec{v}$ in the background of electric field $\vec{\mathbf{E}}$ and magnetic field $\vec{\mathbf{B}}$ experiences force

$$\vec{F} = q(\vec{\mathbf{E}} + \vec{v} \times \vec{\mathbf{B}}).$$

This is the Lorentz force law. We can talk about force density $\vec{f}$ per volume. Then the Lorentz force law becomes

$$f = \rho \vec{\mathbf{E}} + \vec{\mathbf{j}} \times \vec{\mathbf{B}}.$$

Maxwell's equations, together with the Lorentz force law, form the foundation of classical electromagnetism.

Maxwell's equations have a compact form in geometric terms. We collect electric and magnetic field into a 2-form F on the space-time $\mathbb{R}^{3,1}$ as

$$F = (\mathbf{E}_x dx + \mathbf{E}_y dy + \mathbf{E}_z dz) \wedge dt + \mathbf{B}_x dy \wedge dz + \mathbf{B}_y dz \wedge dx + \mathbf{B}_z dx \wedge dy.$$

Taking the de Rham differential, we find

$$\begin{aligned} dF = dt \wedge [(\partial_x \mathbf{E}_y - \partial_y \mathbf{E}_x) dx \wedge dy + (\partial_y \mathbf{E}_z - \partial_z \mathbf{E}_y) dy \wedge dz + (\partial_z \mathbf{E}_x - \partial_x \mathbf{E}_z) dz \wedge dx] \\ + dt \wedge (\partial_t \mathbf{B}_x dy \wedge dz + \partial_t \mathbf{B}_y dz \wedge dx + \partial_t \mathbf{B}_z dx \wedge dy) \\ + (\partial_x \mathbf{B}_x + \partial_y \mathbf{B}_y + \partial_z \mathbf{B}_z) dx \wedge dy \wedge dz. \end{aligned}$$

We observe that two of Maxwell's equations can be equivalently described as

$$\begin{cases} \nabla \cdot \vec{\mathbf{B}} = 0 \\ \nabla \times \vec{\mathbf{E}} = -\frac{\partial \vec{\mathbf{B}}}{\partial t} \end{cases} \qquad \Longleftrightarrow \qquad dF = 0.$$

In other words, F is a closed 2-form.

Let us consider the Hodge dual 2-form

$$*F = \frac{1}{c}(\mathbf{E}_x dy \wedge dz + \mathbf{E}_y dz \wedge dx + \mathbf{E}_z dx \wedge dy) - c(\mathbf{B}_x dx + \mathbf{B}_y dy + \mathbf{B}_z dz) \wedge dt.$$

We see that $*$ switches electric and magnetic field

$$* : \vec{\mathbf{E}} \longmapsto -c\vec{\mathbf{B}}$$
$$* : \vec{\mathbf{B}} \longmapsto \frac{1}{c}\vec{\mathbf{E}}$$

It is computed

$$\begin{aligned} *d*F = &\frac{1}{c^2}(\partial_t \mathbf{E}_x dx + \partial_t \mathbf{E}_y dy + \partial_t \mathbf{E}_z dz) + (\partial_x \mathbf{E}_x + \partial_y \mathbf{E}_y + \partial_z \mathbf{E}_z)dt \\ &- (\partial_x \mathbf{B}_y - \partial_y \mathbf{B}_x)dz - (\partial_y \mathbf{B}_z - \partial_z \mathbf{B}_y)dx - (\partial_z \mathbf{B}_x - \partial_x \mathbf{B}_z)dy. \end{aligned}$$

Let us define the current 1-form

$$J = \rho/\varepsilon_0\, dt - \mu_0(\mathbf{j}_x dx + \mathbf{j}_y dy + \mathbf{j}_z dz).$$

Then the other two Maxwell's equations becomes

$$\begin{cases} \nabla \cdot \vec{\mathbf{E}} = \rho/\varepsilon_0 \\ \nabla \times \vec{\mathbf{B}} = \mu_0 \left(\vec{\mathbf{j}} + \varepsilon_0 \frac{\partial \vec{\mathbf{E}}}{\partial t}\right) \end{cases} \qquad \Longleftrightarrow \qquad d^*F = J.$$

Here $d^* = *d*$ is the adjoint of d. We have arrived at the geometric form of Maxwell's equations

$$\begin{cases} dF = 0 \\ d^*F = J \end{cases}$$

1.3 Conservation Law of Charge

Electric Charge

The words "electric" and "electricity" come from the Greek word for "amber". Electric charge is an intrinsic property of matter, and all take value in an integral[1] multiple of the elementary charge

$$e = 1.602176634 \times 10^{-19}\, C.$$

A single proton carries electric charge e and a single electron carries electric charge $-e$.

[1]Here we do not consider quarks or quasi-particles.

Since e is practically small, it is natural to consider continuous objects and define "charge density" $\rho(\vec{r}, t)$ per unit volume. Here

$$\vec{r} = (x, y, z)$$

refers to the space coordinate. The total charge Q in a finite volume V is

$$Q = \int_V d^3r\, \rho$$

where $d^3r = dxdydz$ is the standard volume form on the space.

When dealing with point charges q_α located at position $\vec{r}_\alpha$, we can use δ-function and represent the charge density as

$$\rho = \sum_\alpha q_\alpha\, \delta(\vec{r} - \vec{r}_\alpha).$$

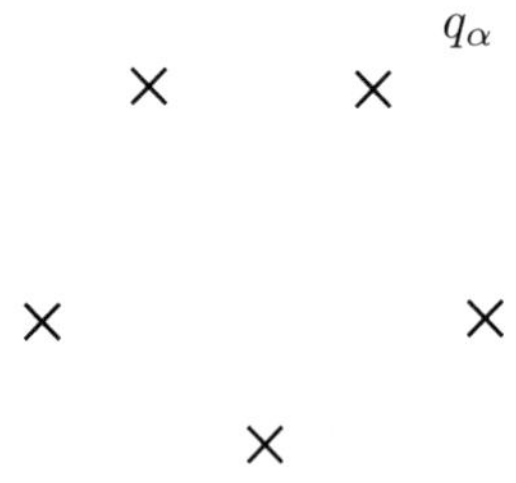

There is a similar treatment for line charges (using δ-function supported on the line)

or surface charges (using δ-function supported on the surface)

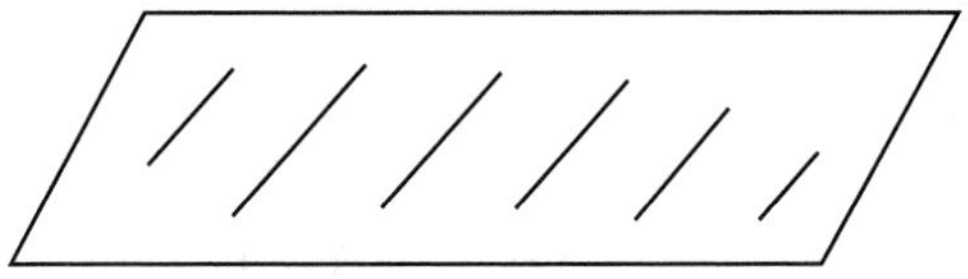

Electric Current

The movement of electric charges constitutes the "electric current". Such quantity is captured by a space-vector $\vec{\mathbf{j}}$ called current density. In general $\vec{\mathbf{j}} = \vec{\mathbf{j}}(\vec{r}, t)$ depends on the position and may change with time. For any surface S, the surface integral

$$I = \int_S d\vec{\sigma} \cdot \vec{\mathbf{j}}$$

counts the charge per unit time passing through S.

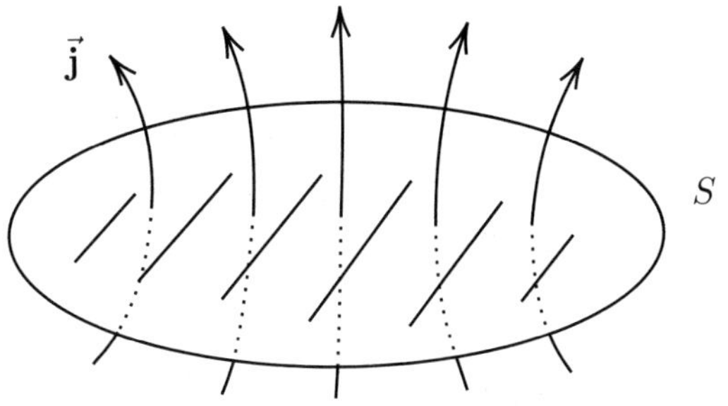

If we consider a charge distribution ρ in which the velocity of a small volume at point $\vec{r}$ is $\vec{v} = \vec{v}(\vec{r}, t)$, then the current density is

$$\vec{\mathbf{j}} = \rho \vec{v}.$$

For example, the current density of a point particle of charge q moving at position $\vec{r}_\alpha(t)$ is

$$\vec{\mathbf{j}} = q\,\dot{\vec{r}}_\alpha(t)\,\delta(\vec{r} - \vec{r}_\alpha(t)).$$

Conservation Law of Charge

Electric charge is conserved in physical processes. Consider the total charge Q contained in some fixed region V

$$Q = \int_V d^3r\,\rho.$$

Its change with time is given by

$$\frac{dQ}{dt} = \int_V d^3r\,\frac{\partial \rho}{\partial t}.$$

On the other hand, we can compute its change by counting the flow out through its boundary $S = \partial V$

$$\frac{dQ}{dt} = -\int_S d\vec{\sigma} \cdot \vec{\mathbf{j}} \,\underset{\text{Theorem}}{\overset{\text{Gauss's}}{=\!=\!=}}\, -\int_V d^3r\,\nabla \cdot \vec{\mathbf{j}}.$$

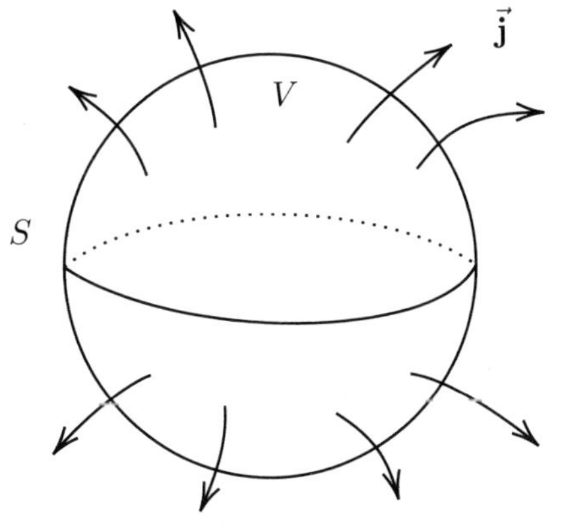

By comparing the two expressions, we find

$$\int_V d^3r \left(\frac{\partial \rho}{\partial t} + \nabla \cdot \vec{\mathbf{j}}\right) = 0.$$

This holds for any region V, leading to the following local form of conservation law of charge

$$\frac{\partial \rho}{\partial t} + \nabla \cdot \vec{\mathbf{j}} = 0$$

This equation is also a consistency equation for Maxwell's equations. To see this, recall

$$d^*F = J$$

where

$$J = \frac{1}{\varepsilon_0}\left(\rho dt - \frac{1}{c^2}(\mathbf{j}_x dx + \mathbf{j}_y dy + \mathbf{j}_z dz)\right).$$

The equation $d^*F = J$ is equivalent to

$$d(*F) = *J.$$

Since $d^2 = 0$, the consistency of this equation requires

$$d(*J) = 0.$$

Explicitly, we have

$$*J = \frac{1}{c\varepsilon_0}(\rho dx \wedge dy \wedge dz - \mathbf{j}_x dt \wedge dy \wedge dz - \mathbf{j}_y dt \wedge dz \wedge dx - \mathbf{j}_z dt \wedge dx \wedge dy),$$

$$d(*J) = \frac{1}{c\varepsilon_0}\left(\frac{\partial \rho}{\partial t} + \nabla \cdot \vec{\mathbf{j}}\right) dt \wedge dx \wedge dy \wedge dz.$$

Therefore

$$d(*J) = 0 \qquad \Longleftrightarrow \qquad \frac{\partial \rho}{\partial t} + \nabla \cdot \vec{\mathbf{j}} = 0.$$

The conservation law can be also understood via Stokes' Theorem as follows. Consider the following region

$$M = V \times [t_1, t_2] \quad \text{in the spacetime } \mathbb{R}^{3,1}.$$

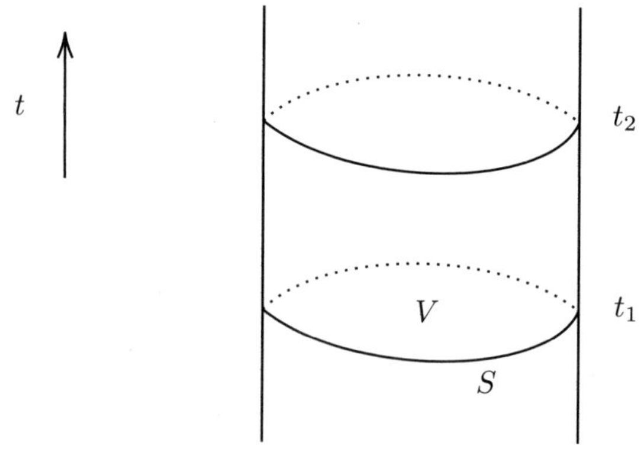

The boundary ∂M of M consists of three pieces

$$V \times \{t_2\}, \qquad V \times \{t_1\}, \qquad S \times [t_1, t_2].$$

Applying Stokes' Theorem,

$$0 = c\varepsilon_0 \int_M d(*J) = c\varepsilon_0 \int_{\partial M} *J = \int_V d^3 r \rho(\vec{r}, t_2) - \int_V d^3 r \rho(\vec{r}, t_1) + \int_{t_1}^{t_2} dt \int_S d\vec{\sigma} \cdot \vec{\mathbf{j}}.$$

Taking the infinitesimal form $t_2 \to t_1$ (or simply taking the derivative of t_2), we find

$$\int_V d^3 r \frac{\partial \rho}{\partial t} + \int_S d\vec{\sigma} \cdot \vec{\mathbf{j}} = 0.$$

In Section 4.6.4, we will discuss another interpretation of charge conservation from the point of view of gauge principle.

1.4 Coulomb's Law

In the early 1770s, Henry Cavendish discovered the dependence of the force between charged bodies in an unpublished note. It was later published and fully established by Charles-Augustin de Coulomb and now coined the name "Coulomb's Law": the magnitude of the electric force between two point charges is proportional to the product of the charges and inverse proportional to the square of their distance. Precisely, if we place two point particles of charge q and q' at position $\vec{r}$ and $\vec{r}\,'$, then the particle q will experience a force by

$$\vec{F} = \frac{1}{4\pi\varepsilon_0} q q' \frac{\vec{r} - \vec{r}\,'}{|\vec{r} - \vec{r}\,'|^3}.$$

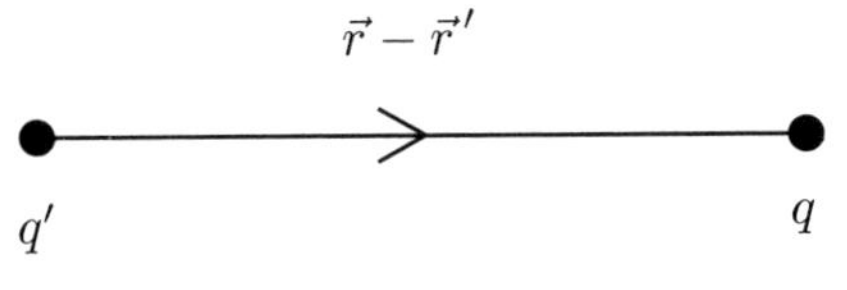

In particular, it is repulsive/attractive if q and q' have the same/opposite signs. If we place several particles $q_1', q_2', \cdots$, then the principle of superposition applies and we get

$$\vec{F} = \frac{1}{4\pi\varepsilon_0} \sum_\alpha q q_\alpha' \frac{\vec{r} - \vec{r}_\alpha\,'}{|\vec{r} - \vec{r}_\alpha\,'|^3}.$$

This formula naturally generalizes to the case with electric charge distribution ρ

$$\vec{F} = \frac{1}{4\pi\varepsilon_0} q \int d^3 r' \, \rho(\vec{r}\,') \frac{\vec{r} - \vec{r}\,'}{|\vec{r} - \vec{r}\,'|^3}.$$

We can recast this formula into the form

$$\vec{F} = q\vec{\mathbf{E}}$$

where the vector $\vec{\mathbf{E}}$ is the electric field produced by the charge distribution

$$\vec{\mathbf{E}}(\vec{r}) = \frac{1}{4\pi\varepsilon_0}\int d^3r'\,\rho(\vec{r}\,')\frac{\vec{r}-\vec{r}\,'}{|\vec{r}-\vec{r}\,'|^3}.$$

Here now comes the important idea and concept of field. The Coulomb's law in the form $\vec{F} = q\vec{\mathbf{E}}$ is a conceptual shift of picture on the nature of force. A particle experiences a force determined by the local value of the field at the position of the particle. This is different from the non-local force over large distance. In particular, fields are physical, and can exist independently of the presence of charged particles.

Now we look into the expression

$$\vec{\mathbf{E}}(\vec{r}) = \frac{1}{4\pi\varepsilon_0}\int d^3r'\,\rho(\vec{r}\,')\frac{\vec{r}-\vec{r}\,'}{|\vec{r}-\vec{r}\,'|^3} = -\frac{1}{4\pi\varepsilon_0}\int d^3r'\,\rho(\vec{r}\,')\nabla\frac{1}{|\vec{r}-\vec{r}\,'|}.$$

Here $\nabla = \left(\frac{\partial}{\partial x}, \frac{\partial}{\partial y}, \frac{\partial}{\partial z}\right)$ is the gradient operator with respect to the position $\vec{r}$. Observe that $\frac{1}{|\vec{r}-\vec{r}\,'|}$ is the Green's function on $\mathbb{R}^3$, or precisely

$$\nabla^2\frac{1}{|\vec{r}-\vec{r}\,'|} = -4\pi\delta(\vec{r}-\vec{r}\,').$$

Here $\nabla^2 = \partial_x^2 + \partial_y^2 + \partial_z^2$ is the Laplacian operator on $\mathbb{R}^3$. It follows that

$$\nabla\cdot\vec{\mathbf{E}}(\vec{r}) = -\frac{1}{4\pi\varepsilon_0}\int d^3r'\,\rho(\vec{r}\,')\nabla^2\frac{1}{|\vec{r}-\vec{r}\,'|} = \frac{1}{\varepsilon_0}\int d^3r'\,\rho(\vec{r}\,')\delta(\vec{r}-\vec{r}\,') = \rho(\vec{r})/\varepsilon_0,$$

i.e.

$$\nabla\cdot\vec{\mathbf{E}} = \rho/\varepsilon_0.$$

We have found that Coulomb's law gives one of the four Maxwell's equations.

Gauss's Law

The equation $\nabla\cdot\vec{\mathbf{E}} = \rho/\varepsilon_0$ has the interpretation that the electric field is sourced by the electric charge. To illustrate this, consider some region V with boundary $\partial V = S$. We consider the electric flux through S defined by the surface integral

$$\int_S d\vec{\sigma}\cdot\vec{\mathbf{E}}.$$

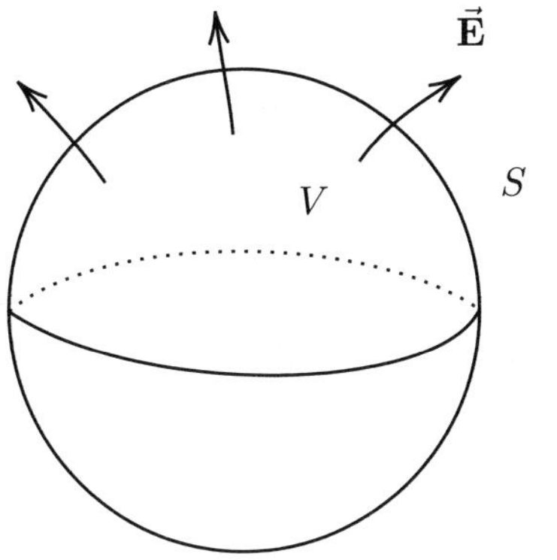

By Gauss's Theorem, this is equal to

$$\int_V d^3r\, \nabla \cdot \vec{\mathbf{E}} = \frac{1}{\varepsilon_0} \int_V d^3r\, \rho = \frac{Q_V}{\varepsilon_0}$$

where $Q_V = \int_V d^3r\, \rho$ is the total charge inside V. This is Gauss's law. In particular, let us assume the electric charge is supported in some region N, and consider a surface S surrounding the entire N as pictured.

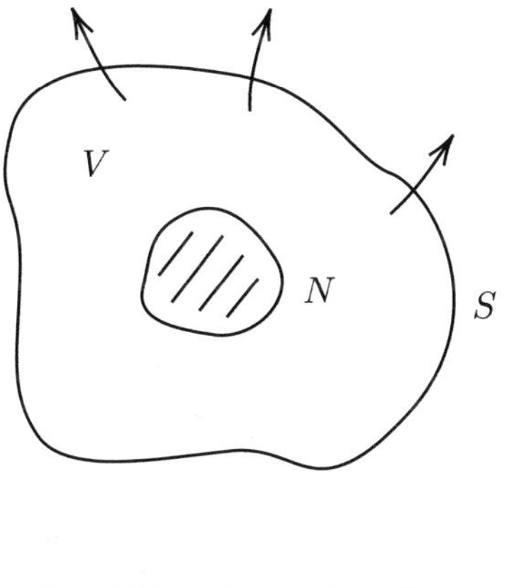

In other words, $N \subset V$, $S = \partial V$. Then the surface flux

$$\int_S d\vec{\sigma} \cdot \vec{\mathbf{E}} = \frac{1}{\varepsilon_0} \int_N d^3r\, \rho$$

is always the same: it does not matter what shape the surface S takes, as long as it surrounds the entire N

$$\int_S d\vec{\sigma} \cdot \vec{\mathbf{E}} = \int_{S'} d\vec{\sigma} \cdot \vec{\mathbf{E}}.$$

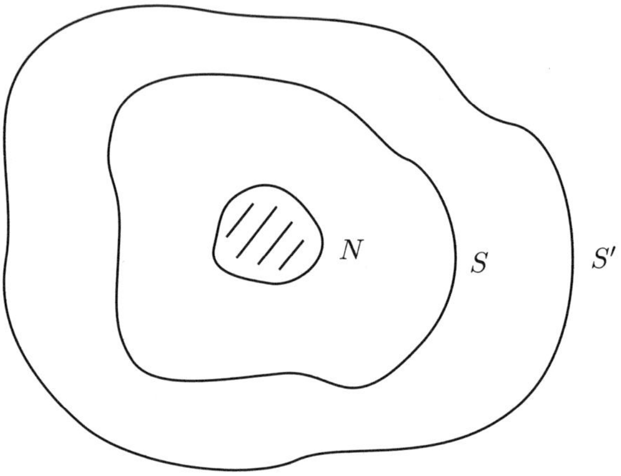

Geometrically, we have

$$d(*F) = *J.$$

Away from the region with electric charge and current,

$$d(*F) = 0.$$

Therefore Stokes' Theorem tells that inside those regions

$$\int_S *F = \frac{1}{c}\int_S \mathbf{E}_x dy \wedge dz + \mathbf{E}_y dz \wedge dx + \mathbf{E}_z dx \wedge dy = \frac{1}{c}\int_S d\vec{\sigma}\cdot\vec{\mathbf{E}}$$

is invariant under continuous deformations of the surface S.

1.5 Ampère's Law

As we have seen,

charge ρ	$\Longrightarrow$	electric field $\vec{\mathbf{E}}$

In 1800, Alessandro Volta invented the Voltaic pile which produces a steady electric current. This can be viewed as an early electric battery, and enabled a rapid many discoveries in chemistry and electromagnetism. The question of a possible interaction between electricity and magnetism had arisen soon after the invention of Volta's pile. In 1820, Hans Christian Ørsted discovered that a compass needle was deflated from magnetic north by a nearby electric current, confirming the first connection between electricity and magnetism. This phenomenon can be summarized as

current j	$\Longrightarrow$	magnetic field $\vec{\mathbf{B}}$

Soon after Ørsted's discovery, André-Marie Ampère found that two parallel wires carrying electric currents attract or repel each other. Such mutual action was formulated in mathematics and led to the important principal: Ampère's law. In 1827,

Ampère published his book *Memoir on the Mathematical Theory of Electrodynamic Phenomena, Uniquely Deduced from Experience*, which coined the name "electrodynamics".

Let us first consider steady solution of Maxwell's equation, i.e. all things are independent of time t. The equations governing the magnetic fields in the steady case are

$$\begin{cases} \nabla \times \vec{\mathbf{B}} = \mu_0 \vec{\mathbf{j}} & \text{Ampère's law} \\ \nabla \cdot \vec{\mathbf{B}} = 0 & \text{Gauss's law} \end{cases}$$

In the steady situation, $\frac{\partial \rho}{\partial t} = 0$ and hence the charge conservation law becomes

$$\nabla \cdot \vec{\mathbf{j}} = 0.$$

This is compatible with Ampère's Law since

$$\nabla \cdot \left(\nabla \times \vec{\mathbf{B}}\right) = 0 \qquad \text{for any vector } \vec{\mathbf{B}}.$$

We will next give a geometric description of the above two equations and their solutions.

1.6 Biot-Savart Law

In 1820 after Ørsted's discovery, Jean-Baptiste Biot and Félix Savart discovered an equation describing the magnetic field generated by a constant electric current. This is now called Biot-Savart law, which is a fundamental law in magnetostatics.

Let us come back to the geometry of the following Maxwell's equation for magnetic fields in the steady situation (magnetostatics)

$$\begin{cases} \nabla \times \vec{\mathbf{B}} = \mu_0 \vec{\mathbf{j}} \\ \nabla \cdot \vec{\mathbf{B}} = 0 \end{cases}$$

Let us recollect the vectors $\vec{\mathbf{E}} = (\mathbf{E}_x, \mathbf{E}_y, \mathbf{E}_z)$, $\vec{\mathbf{B}} = (\mathbf{B}_x, \mathbf{B}_y, \mathbf{B}_z)$ and $\vec{\mathbf{j}} = (\mathbf{j}_x, \mathbf{j}_y, \mathbf{j}_z)$ as 1-forms on $\mathbb{R}^3$

$$\begin{cases} \mathbb{E} = \mathbf{E}_x dx + \mathbf{E}_y dy + \mathbf{E}_z dz \\ \mathbb{B} = \mathbf{B}_x dx + \mathbf{B}_y dy + \mathbf{B}_z dz \\ \mathbb{j} \;= \mathbf{j}_x dx + \mathbf{j}_y dy + \mathbf{j}_z dz \end{cases}$$

Then it is easy to see that the vector $\nabla \times \vec{\mathbf{B}}$ corresponds to the 1-form $*_3 d_3 \mathbb{B}$ and

$$\nabla \cdot \vec{\mathbf{B}} = -d_3^* \mathbb{B}.$$

Therefore we have the following geometric form (using $(*_3)^2 = 1$)

$$\begin{cases} \nabla \times \vec{\mathbf{B}} = \mu_0 \vec{\mathbf{j}} \\ \nabla \cdot \vec{\mathbf{B}} = 0 \end{cases} \iff \begin{cases} d_3 \mathbb{B} = \mu_0 *_3 \mathbb{j} \\ d_3^* \mathbb{B} = 0 \end{cases}$$

Here recall d_3 deontes the de Rham differential on $\mathbb{R}^3$. The consistency condition $d_3^2\mathbb{B} = 0$ asks for $d_3 *_3 \mathbb{j} = 0$, or equivalently $\nabla \cdot \vec{\mathbf{j}} = 0$. This is the steady charge conservation.

From $d_3^*\mathbb{B} = 0$ or equivalently $d_3 *_3 \mathbb{B} = 0$, we can write $*_3\mathbb{B} = d_3\mathbb{A}$ for a 1-form

$$\mathbb{A} = \mathbf{A}_x dx + \mathbf{A}_y dy + \mathbf{A}_z dz.$$

Such a vector $\vec{\mathbf{A}} = (\mathbf{A}_x, \mathbf{A}_y, \mathbf{A}_z)$ is called the "vector potential", and will be playing an important role in electromagnetism. In vector notation, this is equivalent to the equation

$$\vec{\mathbf{B}} = \nabla \times \vec{\mathbf{A}}.$$

The vector potential is not uniquely specified. Indeed, for any function χ, the shift

$$\mathbb{A} \longmapsto \mathbb{A} + d_3\chi = \mathbb{A}'$$

also satisfies

$$d_3\mathbb{A}' = d_3(\mathbb{A} + d_3\chi) = d_3\mathbb{A} = *_3\mathbb{B}$$

since $d_3^2 = 0$. Such a change of $\mathbb{A}$ is called a gauge transformation, which we will explain in detail in later part of the note.

In turns out that there is a condition to fix this freedom of choice, by asking $\mathbb{A}$ to satisfy the "Coulomb gauge" condition

$$d_3^*\mathbb{A} = 0, \qquad \text{or in vector notation,} \qquad \nabla \cdot \vec{\mathbf{A}} = 0.$$

Let us now assume $\mathbb{A}$ satisfies the Coulomb gauge condition. Let us substitute $\mathbb{A}$ into

$$d_3(*_3 d_3\mathbb{A}) = \mu_0 *_3 \mathbb{j}.$$

This equation is the same as

$$d_3^* d_3\mathbb{A} = \mu_0\mathbb{j}.$$

Coulomb gauge condition implies

$$d_3 d_3^*\mathbb{A} = 0.$$

Combining the above two and using $d_3 d_3^* + d_3^* d_3 = -\nabla^2$, we find

$$\nabla^2\mathbb{A} = -\mu_0\mathbb{j}.$$

This can be solved using Green's function

$$\vec{\mathbf{A}}(\vec{r}) = \frac{\mu_0}{4\pi}\int d^3r' \, \frac{\vec{\mathbf{j}}(\vec{r}\,')}{|\vec{r} - \vec{r}\,'|}.$$

It can be checked that $\vec{\mathbf{A}}(\vec{r})$ indeed satisfies the Coulomb gauge condition using $\nabla \cdot \vec{\mathbf{j}} = 0$ (Exercise). Now we can write down the magnetic field $\vec{\mathbf{B}}$ in the presence of steady current $\vec{\mathbf{j}}$ by

$$\vec{\mathbf{B}}(\vec{r}) = \nabla \times \vec{\mathbf{A}}(\vec{r}) = \frac{\mu_0}{4\pi}\int d^3r' \, \nabla\left(\frac{1}{|\vec{r} - \vec{r}\,'|}\right) \times \vec{\mathbf{j}}(\vec{r}\,'),$$

i.e.

$$\vec{\mathbf{B}}(\vec{r}) = \frac{\mu_0}{4\pi} \int d^3r' \, \vec{\mathbf{j}}(\vec{r}\,') \times \frac{\vec{r} - \vec{r}\,'}{|\vec{r} - \vec{r}\,'|^3}.$$

This formula is known as **Biot-Savart Law**.

1.7 Gauss's Linking Formula

Consider two closed, non-intersecting curves C_1 and C_2 in $\mathbb{R}^3$. There is an integer $n(C_1, C_2)$, called the **linking number** of C_1 and C_2, which describes how many times one of the curve winds around the other.

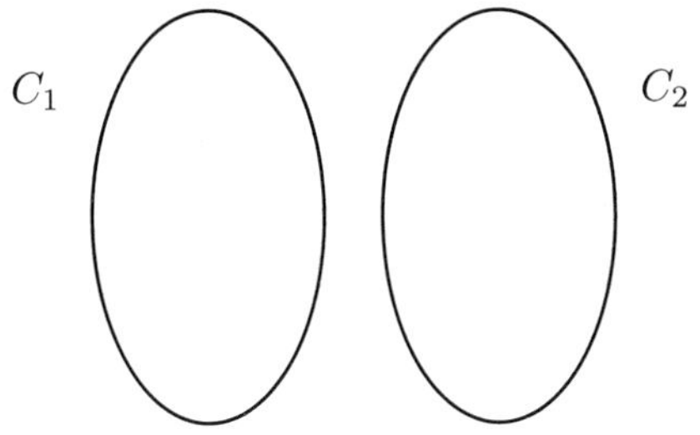

$n(C_1, C_2) = 0$

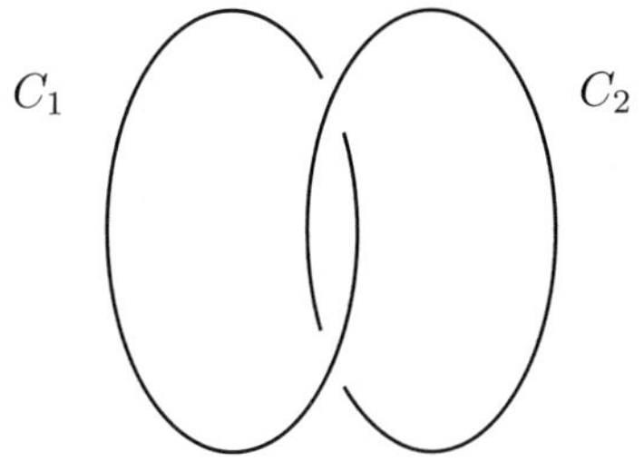

$n(C_1, C_2) = \pm 1$

The sign depends on the orientation of C_1 and C_2.

One way to define the linking number in the current situation is as follows. We first fill C_1 by a disk D_1 whose boundary is precisely $\partial D_1 = C_1$. Then the linking number

$$n(C_1, C_2) = \#(D_1 \cap C_2)$$

counts (with sign) the number of intersection of D_1 with C_2.

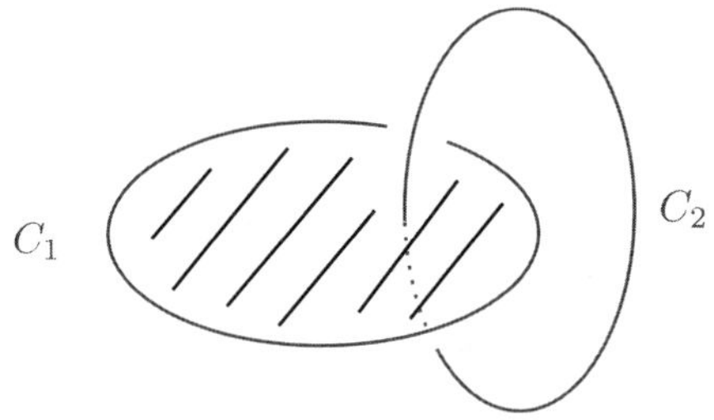

$n(C_1, C_2) = \pm 1$

You can similarly fill C_2 first by a disk and intersect with C_1: the answer is the same.

There is an integral expression for the linking number due to Gauss, which is closely related to ingredients of electromagnetism. Suppose C_2 carries a steady current I. It creates a magnetic field $\vec{\mathbf{B}}$ in space, which can be computed via Biot-Savart formula

$$\vec{\mathbf{B}}(\vec{r}_1) = \frac{\mu_0 I}{4\pi} \oint_{C_2} d\vec{r}_2 \times \frac{\vec{r}_1 - \vec{r}_2}{|\vec{r}_1 - \vec{r}_2|^3}.$$

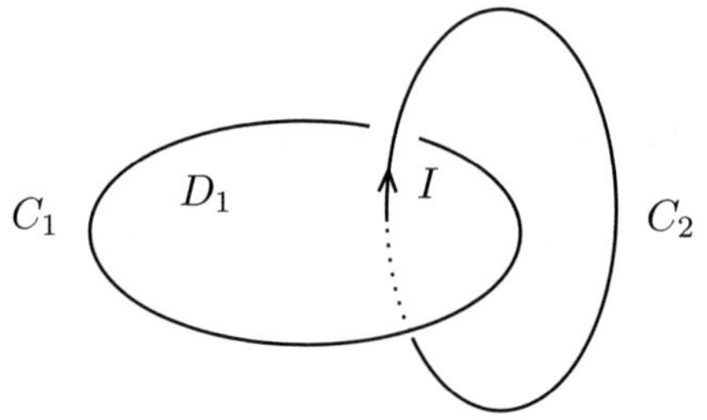

Let us consider the circle integral of $\vec{\mathbf{B}}$ along C_1

$$\oint_{C_1} d\vec{r}_1 \cdot \vec{\mathbf{B}}(\vec{r}_1) = \frac{\mu_0 I}{4\pi} \oint_{C_1} d\vec{r}_1 \cdot \oint_{C_2} d\vec{r}_2 \times \frac{\vec{r}_1 - \vec{r}_2}{|\vec{r}_1 - \vec{r}_2|^3}.$$

On the other hand, by Stokes' Theorem and Ampère's Law

$$\oint_{C_1} d\vec{r}_1 \cdot \vec{\mathbf{B}}(\vec{r}_1) = \int_{D_1} d\vec{\sigma} \cdot (\nabla \times \vec{\mathbf{B}}) = \mu_0 \int_{D_1} d\vec{\sigma} \cdot \vec{\mathbf{j}} = \mu_0 I\, n(C_1, C_2).$$

Comparing the above two expressions, we find

$$n(C_1, C_2) = \frac{1}{4\pi} \oint_{C_1} d\vec{r}_1 \cdot \oint_{C_2} d\vec{r}_2 \times \frac{\vec{r}_1 - \vec{r}_2}{|\vec{r}_1 - \vec{r}_2|^3} = \frac{1}{4\pi} \oint_{C_1} \cdot \oint_{C_2} \frac{(\vec{r}_1 - \vec{r}_2) \cdot (d\vec{r}_1 \times d\vec{r}_2)}{|\vec{r}_1 - \vec{r}_2|^3}.$$

This is precisely the linking number formula of Gauss.

1.8 Faraday's Law

In 1831, Michael Faraday discovered electromagnetic induction, which shows that the change of magnetic field produces electric field. This is now called **Faraday's Law of Induction**. Faraday explained electromagnetic induction using a concept called lines of force, and essentially proposed the concept of electromagnetic field.

Faraday's law says

change of magnetic field $\Longrightarrow$ electric field

Precisely, this is described by the third Maxwell's equation

$$\nabla \times \vec{\mathbf{E}} = -\frac{\partial \vec{\mathbf{B}}}{\partial t}.$$

In terms of differential forms as described in Section 1.6, this is written as

$$*_3 d_3 \mathbb{E} = -\frac{\partial \mathbb{B}}{\partial t}$$

or equivalently

$$d_3\mathbb{E} = -\frac{\partial}{\partial t}(*_3\mathbb{B}).$$

Let us consider a surface S which is bounded by a closed curve C.

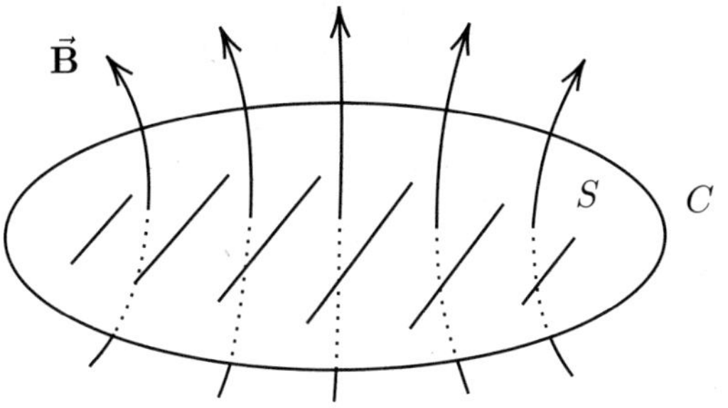

The integral of the magnetic field over the surface S is called the "**magnetic flux**" through S

$$\Phi = \int_S d\vec{\sigma} \cdot \vec{\mathbf{B}} = \int_S *_3\mathbb{B}.$$

Then we have the integral form of Faraday's law by

$$-\frac{d}{dt}\int_S *_3\mathbb{B} = \int_S \left(-\frac{\partial}{\partial t} *_3\mathbb{B}\right) = \int_S d_3\mathbb{E} = \int_C \mathbb{E},$$

i.e.

$$-\frac{d}{dt}\int_S d\vec{\sigma} \cdot \vec{\mathbf{B}} = \int_C d\vec{r} \cdot \vec{\mathbf{E}}.$$

This is the form when Faraday discovered the law of induction: changing the magnetic flux through S will produce a current to flow along C.

1.9 Ampère-Maxwell Law

In Magnerostatics for the steady situation, we have Ampère's law on the magnetic field arising from electric current

$$\nabla \times \vec{\mathbf{B}} = \mu_0 \vec{\mathbf{j}}.$$

However, in the dynamical case when all things depend on time, this is not enough. Faraday's law says that the change of magnetic field will induce electric field. Parallelly, it is natural to ask whether the change of electric field will induce magnetic field. This is Maxwell's addition to Ampère's Law, usually called Ampère-Maxwell Law

$$\nabla \times \vec{\mathbf{B}} = \mu_0 \left(\vec{\mathbf{j}} + \varepsilon_0 \frac{\partial \vec{\mathbf{E}}}{\partial t}\right),$$

or using the relation $\mu_0\varepsilon_0 = \frac{1}{c^2}$,

$$\nabla \times \vec{\mathbf{B}} = \mu_0 \vec{\mathbf{j}} + \frac{1}{c^2}\frac{\partial \vec{\mathbf{E}}}{\partial t}.$$

The extra term is called "**displacement current**", though it is not really a current but an addition to the current. In terms of differential forms, Ampère-Maxwell Law is

$$d_3\mathbb{B} = \mu_0 {*_3}\mathbb{j} + \frac{1}{c^2}\frac{\partial}{\partial t}{*_3}\mathbb{E}.$$

The displacement current is in fact necessary for the consistency with charge conservation. In fact, using $(d_3)^2 = 0$

$$0 = d_3 d_3 \mathbb{B} = \mu_0 d_3 {*_3}\mathbb{j} + \frac{1}{c^2} d_3 {*_3}\frac{\partial}{\partial t}\mathbb{E} = \left(\mu_0 \nabla \cdot \vec{\mathbf{j}} + \frac{1}{c^2}\frac{\partial}{\partial t}\nabla \cdot \vec{\mathbf{E}}\right) dx \wedge dy \wedge dz,$$

from which we get

$$\mu_0 \nabla \cdot \vec{\mathbf{j}} + \frac{1}{c^2}\frac{\partial}{\partial t}\nabla \cdot \vec{\mathbf{E}} = 0.$$

This is precisely the conservation law of charge using $\nabla \cdot \vec{\mathbf{E}} = \rho/\varepsilon_0$.

1.10 Potential and Gauge

We have discussed the full set of Maxwell's equations

$$\begin{cases} \nabla \cdot \vec{\mathbf{E}} = \rho/\varepsilon_0 \\ \nabla \cdot \vec{\mathbf{B}} = 0 \\ \nabla \times \vec{\mathbf{E}} = -\frac{\partial \vec{\mathbf{B}}}{\partial t} \\ \nabla \times \vec{\mathbf{B}} = \mu_0 \left(\vec{\mathbf{j}} + \varepsilon_0 \frac{\partial \vec{\mathbf{E}}}{\partial t}\right) \end{cases}$$

and its geometrical form

$$\begin{cases} dF = 0 \\ d(*F) = *J \end{cases}$$

where $F = \mathbb{E} \wedge dt + {*_3}\mathbb{B}$ is the electro-magnetic field and $J = \rho/\varepsilon_0 dt - \mu_0 \mathbb{j}$ is the charge current.

The consistency condition

$$d(*J) = 0$$

is the the conservation law of charge

$$\frac{\partial \rho}{\partial t} + \nabla \cdot \vec{\mathbf{j}} = 0.$$

Consider the equation

$$dF = 0.$$

Since the topology of $\mathbb{R}^{3,1}$ is trivial, $H^2_{\mathrm{dR}}(\mathbb{R}^{3,1}) = 0$. Hence we can always find a 1-form A on $\mathbb{R}^{3,1}$ such that

$$F = dA.$$

In general when spacetime has nontrivial topology, such A can only be obtained locally and glued via transformation law. We will discuss this situation in more detail in Section 4.6.

Let us write the 1-form A as

$$A = -\phi dt + \mathbb{A}$$

where $\mathbb{A} = \mathbf{A}_x dx + \mathbf{A}_y dy + \mathbf{A}_z dz$. Then the relation $F = dA$ becomes

$$\mathbb{E} \wedge dt + *_3\mathbb{B} = -d_3\phi \wedge dt + dt \wedge \partial_t\mathbb{A} + d_3\mathbb{A}.$$

Comparing the form types of both sides, we find

$$\begin{cases} \mathbb{E} = -d_3\phi - \partial_t\mathbb{A} \\ \mathbb{B} = *_3 d_3\mathbb{A} \end{cases}$$

In vector notation, let $\vec{\mathbf{A}} = (\mathbf{A}_x, \mathbf{A}_y, \mathbf{A}_z)$, this is

$$\begin{cases} \vec{\mathbf{E}} = -\nabla\phi - \partial_t\vec{\mathbf{A}} \\ \vec{\mathbf{B}} = \nabla \times \vec{\mathbf{A}} \end{cases}$$

ϕ is called the **scalar potential**. $\vec{\mathbf{A}}$ is the **vector potential** that we have seen in Section 1.6.

Again, the choice of A is not unique. We can always shift it by

$$A \longmapsto A + d\chi$$

for a function $\chi = \chi(\vec{r}, t)$ on the spacetime. The electric-magnetic field remains the same

$$F = dA \longmapsto d(A + d\chi) = dA.$$

These are **gauge transformations**, and A is called the **gauge field**. We will study gauge theory systematically in Chapter 4.

In components, the gauge transformation is

$$\begin{cases} \phi \longmapsto \phi - \dfrac{\partial\chi}{\partial t} \\ \vec{\mathbf{A}} \longmapsto \vec{\mathbf{A}} + \nabla\chi \end{cases}$$

In terms of the gauge field, Maxwell's equations become a single equation

$$*d * dA = J.$$

To solve this equation, we can choose a gauge condition to fix the gauge degree of freedom. There are usually two commonly used gauge fixing condition:

$$\underline{\text{Coulomb gauge}}: \quad \nabla \cdot \vec{\mathbf{A}} = 0$$

$$\underline{\text{Lorenz gauge}}: \quad \nabla \cdot \vec{\mathbf{A}} + \frac{1}{c^2}\frac{\partial \phi}{\partial t} = 0$$

We have seen Coulomb gauge in Section 1.6 on the Biot-Savart Law in magnetostatics. Here we briefly comment on Lorenz gauge which we will discuss later in details.

The Lorenz gauge can be written as

$$\underline{\text{Lorenz gauge}}: \qquad d * A = 0.$$

In Lorenz gauge, we have

$$\begin{cases} d^* dA = J \\ d^* A = 0 \end{cases}$$

So

$$\Box A = -(dd^* + d^* d)A = -J,$$

or in components,

$$\begin{cases} \nabla^2 \phi - \dfrac{1}{c^2}\dfrac{\partial^2}{\partial t^2}\phi = -\rho/\varepsilon_0 \\ \nabla^2 \vec{\mathbf{A}} - \dfrac{1}{c^2}\dfrac{\partial^2}{\partial t^2}\vec{\mathbf{A}} = -\mu_0 \vec{\mathbf{j}} \end{cases}$$

These are inhomogeneous wave equations that we will study in detail in Section 3.4.

In a region without charge and current, such as vacuum, Maxwell's equations reduce to

$$\begin{cases} dF = 0 \\ d^* F = 0 \end{cases}$$

which again leads to the wave equation

$$\Box F = -(dd^* + d^* d)F = 0.$$

In components, we have

$$\begin{cases} \dfrac{1}{c^2}\dfrac{\partial^2 \vec{\mathbf{E}}}{\partial t^2} - \nabla^2 \vec{\mathbf{E}} = 0 \\ \dfrac{1}{c^2}\dfrac{\partial^2 \vec{\mathbf{B}}}{\partial t^2} - \nabla^2 \vec{\mathbf{B}} = 0 \end{cases}$$

which are the standard form of wave equations traveling at the speed of light c. This led Maxwell to propose that light and radio waves were propagating electromagnetic waves. In 1887, Heinrich Hertz demonstrated Maxwell's electromagnetic waves propagating at the same speed as light. This placed Maxwell's theory on a firm foundation.

All electromagnetic waves travel at a fixed speed c in the vaccum, independent of any frame of reference. This looks controversial comparing to classical Newton mechanics. It was studied by Hendrik Lorentz who was able to derive the Lorentz transformations that preserve the form of Maxwell's equations. This subsequently laid the foundation for Einstein's special relativity. We will study this in Chapter 5.

Chapter 2

Static Electromagnetism

In this chapter, we discuss static electromagnetism (including electrostatics and magnetostatics), in which case the electric and magnetic fields do not vary with respect to time.

2.1 Electric Field and Scalar Potential

We still start with the study of electrostatics, which describes the situation of steady charge distribution and no currents. In electrostatics, we have

$$\vec{\mathbf{B}} = \vec{\mathbf{j}} = 0, \qquad \frac{\partial \rho}{\partial t} = 0.$$

The only relevant Maxwell's equations are

$$\begin{cases} \nabla \cdot \vec{\mathbf{E}} = \rho/\varepsilon_0 \\ \nabla \times \vec{\mathbf{E}} = 0 \end{cases}$$

where the electric field $\vec{\mathbf{E}}$ is stationary: $\frac{\partial \vec{\mathbf{E}}}{\partial t} = 0$.

2.1.1 Poisson's and Laplace's Equation

In electrostatics, the potentials are given by

$$\begin{cases} \phi = \phi(\vec{r}) \\ \vec{\mathbf{A}} = 0 \end{cases}$$

In terms of scalar potential, we have

$$\vec{\mathbf{E}} = -\nabla \phi,$$

which automatically solves the equation $\nabla \times \vec{\mathbf{E}} = 0$. Then the equation $\nabla \cdot \vec{\mathbf{E}} = \rho/\varepsilon_0$ becomes

$$\nabla^2 \phi = -\rho/\varepsilon_0.$$

This is **Poisson's equation**.

In the region of space that is free of charge, we have the **Laplace's equation**

$$\nabla^2 \phi = 0.$$

When we consider electrostatics problem in the space without boundary, say $\mathbb{R}^3$, we can solve ϕ (with appropriate decay condition at infinity) in terms of the Coulomb integral

$$\phi(\vec{r}) = \frac{1}{4\pi\varepsilon_0} \int d^3r' \, \frac{\rho(\vec{r}')}{|\vec{r} - \vec{r}'|}.$$

In practice, we usually need to deal with a finite region V in $\mathbb{R}^3$ with boundary $\partial V = S$. The electric field could be stimulated by charges inside or outside V, and we can measure them on the boundary S of V. In this case we need to deal with boundary value problem

$$\begin{cases} \nabla^2 \phi = -\rho/\varepsilon_0 & \text{inside } V \\ \text{boundary condition of } \phi & \text{on } S = \partial V \end{cases}$$

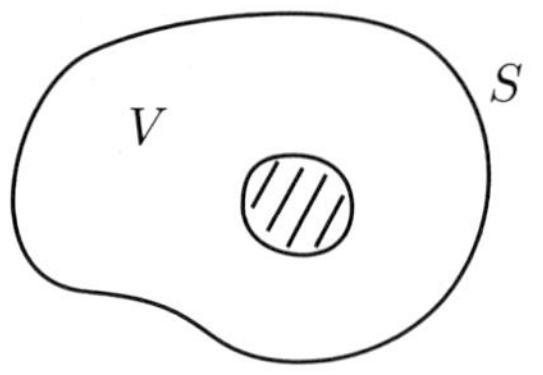

The study of this problem for Poisson's and Laplace's equation is the subject of "**Potential theory**". Let us first look at some examples.

2.1.2 Point Charge

Consider a point charge q located at $\vec{r}_0$. The charge density is

$$\rho(\vec{r}) = q\,\delta(\vec{r} - \vec{r}_0).$$

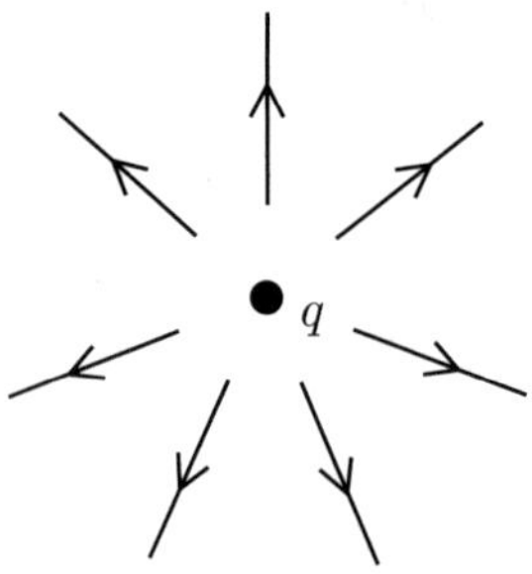

The scalar potential is

$$\phi(\vec{r}) = \frac{1}{4\pi\varepsilon_0} \int d^3r' \, \frac{q\,\delta(\vec{r}-\vec{r}_0)}{|\vec{r}-\vec{r}'|} = \frac{q}{4\pi\varepsilon_0 \, |\vec{r}-\vec{r}_0|}.$$

The electric field is

$$\vec{\mathbf{E}} = -\nabla\phi = \frac{q}{4\pi\varepsilon_0} \frac{\vec{r}-\vec{r}_0}{|\vec{r}-\vec{r}_0|^3},$$

which is the familiar Coulomb's law.

2.1.3 Uniform Ball

Consider a ball B_R of radius R centered at origin, with uniform charge ρ_0 per unit volume. The charge density function in $\mathbb{R}^3$ is

$$\rho = \rho_0 \chi_{B_R}.$$

Here for a subset $A \subset \mathbb{R}^3$, χ_A refers to the characteristic function of A

$$\chi_A(\vec{r}) = \begin{cases} 1 & \vec{r} \in A \\ 0 & \vec{r} \notin A \end{cases}$$

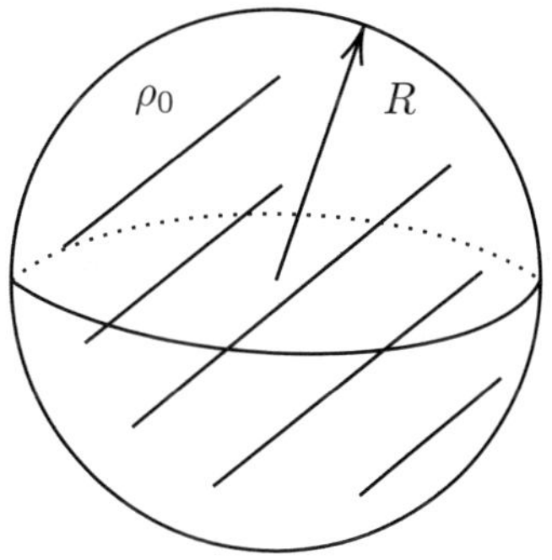

Outside B_R: $|\vec{r}| > R$. The scalar potential is

$$\phi(\vec{r}) = \frac{1}{4\pi\varepsilon_0} \int_{B_R} d^3r' \, \frac{\rho_0}{|\vec{r}-\vec{r}'|}.$$

By spherical symmetry, $\phi = \phi(r)$ is a function of the radius $r = |\vec{r}|$ only. Then

$$\vec{\mathbf{E}} = -\nabla\phi = -\partial_r \phi \frac{\vec{r}}{r}.$$

This is enough to solve $\phi, \vec{\mathbf{E}}$ without explicitly evaluating the above integral. In fact, consider the following surface integral over the sphere S_r of radius r centered at the origin

$$\int_{S_r} d\vec{\sigma} \cdot \nabla\phi = \int_{B_R} d^3r\, \nabla^2\phi = -\int_{B_R} d^3r\, \rho_0/\varepsilon_0 = -\rho_0/\varepsilon_0 \text{ Vol}(B_R) = -\frac{4\pi R^3}{3}\rho_0/\varepsilon_0$$
$$= -Q/\varepsilon_0,$$

where $Q = \rho_0 4\pi R^3/3$ is the total charge of the ball. On the other hand, using $\phi = \phi(r)$,

$$\int_{S_r} d\vec{\sigma} \cdot \nabla\phi = \partial_r\phi \text{ Area}(S_r) = 4\pi r^2 \partial_r\phi.$$

Comparing the above two expressions, we find

$$\partial_r\phi = -\frac{Q}{4\pi\varepsilon_0 r^2}.$$

The electric field outside the sphere is therefore

$$\vec{\mathbf{E}} = -\nabla\phi = \frac{Q}{4\pi\varepsilon_0}\frac{\vec{r}}{r^3} \qquad (r > R)$$

which takes the same form as that of a point charge Q.
Inside B_R: $|\vec{r}| < R$.

We follow the same strategy via spherical symmetry: $\phi = \phi(r)$. Then

$$\int_{S_r} d\vec{\sigma} \cdot \nabla\phi = \int_{B_r} d^3r\, \nabla^2\phi = -\rho_0/\varepsilon_0 \text{ Vol}(B_r) = -\frac{r^3}{R^3}Q/\varepsilon_0,$$

and

$$\int_{S_r} d\vec{S} \cdot \nabla\phi = 4\pi r^2 \partial_r\phi.$$

Comparing these two expressions, we find

$$\partial_r\phi = -\frac{Q}{4\pi\varepsilon_0}\frac{r}{R^3}.$$

The electric field inside the ball is therefore

$$\vec{\mathbf{E}} = -\nabla\phi = \frac{Q}{4\pi\varepsilon_0}\frac{\vec{r}}{R^3} \qquad (r < R)$$

which grows linearly inside the ball. We can also describe a continuous solution of ϕ by

$$\phi(r) = \begin{cases} \dfrac{Q}{4\pi\varepsilon_0}\left(\dfrac{3}{2R} - \dfrac{r^2}{2R^3}\right) & r < R \\ \dfrac{Q}{4\pi\varepsilon_0}\dfrac{1}{r} & r > R \end{cases}$$

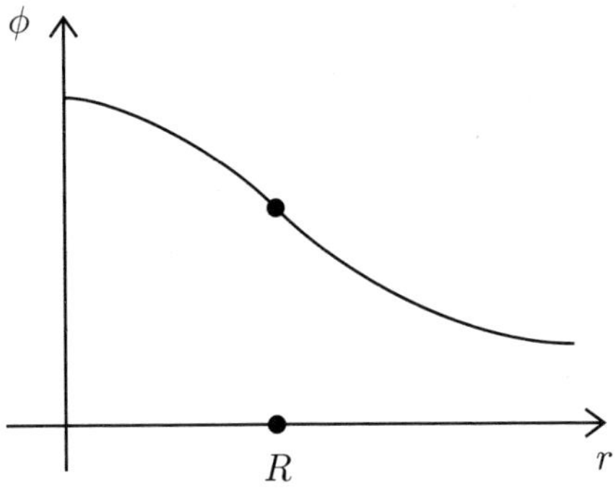

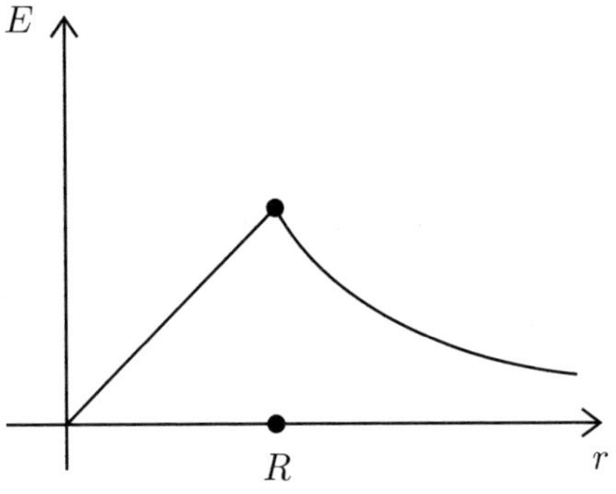

2.1.4 Line Charges

We consider a line segment of length $2L$ with uniform charge λ per unit length, placed along the z-axis and centered at the origin.

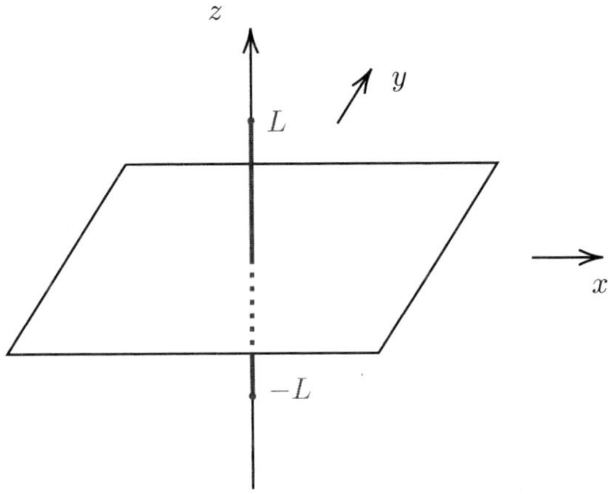

It would be convenient to work with cylindrical coordinates

$$\begin{cases} x = \rho\cos\theta \\ y = \rho\sin\theta \\ z = z \end{cases}$$

The charge density is given by

$$\lambda\,\delta(x)\delta(y)\chi_{|z|\le L}.$$

Due to cylindrical symmetry, the scalar potential is a function of (ρ, z) only and given by

$$\phi(\rho, z) = \frac{\lambda}{4\pi\varepsilon_0}\int_{-L}^{L} dz' \frac{1}{\sqrt{\rho^2 + (z - z')^2}} = \frac{\lambda}{4\pi\varepsilon_0}\ln\left[\frac{\sqrt{(z-L)^2+\rho^2}-(z-L)}{\sqrt{(z+L)^2+\rho^2}-(z+L)}\right].$$

We can simplify this expression using

$$r_\pm = \sqrt{(z \pm L)^2 + \rho^2}$$

which are the distances from the two segment endpoints to the observation point. They are related by

$$r_+^2 - r_-^2 = 4zL.$$

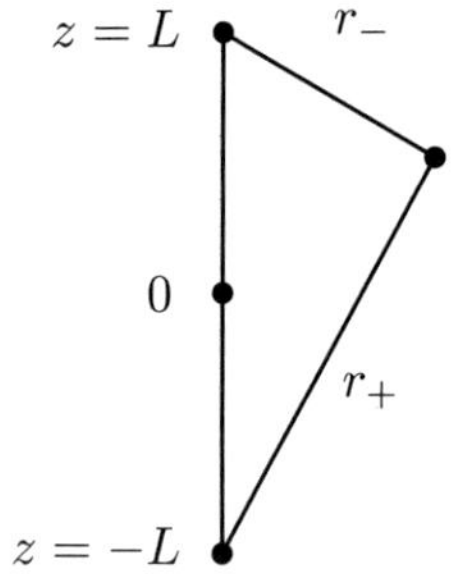

Then

$$\phi = \frac{\lambda}{4\pi\varepsilon_0} \ln\left(\frac{r_- - (z - L)}{r_+ - (z + L)}\right).$$

Observe that

$$\begin{aligned} r_- - (z - L) &= r_- + L - \frac{r_+^2 - r_-^2}{4L} = \frac{(r_- + r_+)(r_- - r_+)}{4L} + r_- + L \\ &= \left(\frac{1}{2}(r_- + r_+) + L\right)\left(\frac{1}{2L}(r_- - r_+) + 1\right). \end{aligned}$$

Similarly,

$$r_+ - (z + L) = \left(\frac{1}{2}(r_- + r_+) - L\right)\left(\frac{1}{2L}(r_- - r_+) + 1\right).$$

It follows that

$$\phi = \frac{\lambda}{4\pi\varepsilon_0} \ln\left(\frac{r_- + r_+ + 2L}{r_- + r_+ - 2L}\right)$$

is a function of $r_- + r_+$ only. This allows us to draw the surface of equipotential, or equivalently

$$r_- + r_+ = \text{constant}$$

which is the geometry of an ellipse. The electric field $\vec{\mathbf{E}} = -\nabla\phi$ is perpendicular to the equipotential surface. This can be visualized by

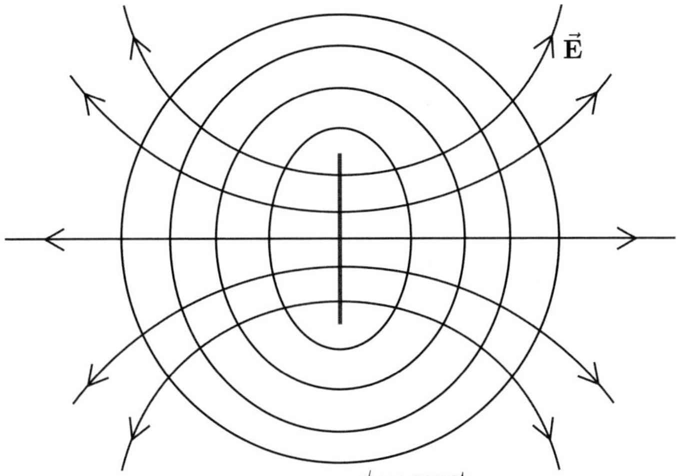

We consider two limit cases.

① $|z| >> L$, $r = \sqrt{\rho^2 + z^2} >> L$.

$$\phi = \frac{\lambda}{4\pi\varepsilon_0} \ln\left(\frac{1 + \frac{2L}{r_+ + r_-}}{1 - \frac{2L}{r_+ + r_-}}\right) \simeq \frac{\lambda}{4\pi\varepsilon_0} \frac{4L}{r_+ + r_-} \simeq \frac{2\lambda L}{4\pi\varepsilon_0 r} = \frac{Q}{4\pi\varepsilon_0 r}.$$

Here $Q = 2\lambda L$ is the total charge. So when we observe very far away the segment, it looks like a point charge as expected.

② $|z| << L$, $\rho << L$.

$$\begin{aligned}\frac{r_+ + r_-}{2L} &= \frac{1}{2L}\left(\sqrt{(L-z)^2 + \rho^2} + \sqrt{(L+z)^2 + \rho^2}\right) \\ &= \frac{1}{2}\left(\sqrt{(1 - z/L)^2 + (\rho/L)^2} + \sqrt{(1 + z/L)^2 + (\rho/L)^2}\right)\end{aligned}$$

Using $\sqrt{1+x} = 1 + \frac{1}{2}x - \frac{1}{8}x^2 + O(x^3)$, the above is equal to

$$\begin{aligned}&= \frac{1}{2}\Bigg(1 + \frac{1}{2}(-2z/L + z^2/L^2 + \rho^2/L^2) - \frac{1}{8}(-2z/L)^2 \\ &\qquad + 1 + \frac{1}{2}(2z/L + z^2/L^2 + \rho^2/L^2) - \frac{1}{8}(-2z/L)^2 + O(L^{-3})\Bigg) \\ &= 1 + \frac{1}{2}(\rho/L)^2 + O(L^{-3}).\end{aligned}$$

Therefore

$$\begin{aligned}\phi &= \frac{\lambda}{4\pi\varepsilon_0}\ln\left(\frac{\frac{r_+ + r_-}{2L}+1}{\frac{r_+ + r_-}{2L}-1}\right) = \frac{\lambda}{4\pi\varepsilon_0}\ln\left(\frac{2+\frac{1}{2}(\rho/L)^2+O(L^{-3})}{\frac{1}{2}(\rho/L)^2+O(L^{-3})}\right)\\ &\simeq \frac{\lambda}{4\pi\varepsilon_0}\ln 4 - \frac{\lambda}{4\pi\varepsilon_0}\ln(\rho/L)^2\\ &= -\frac{\lambda}{2\pi\varepsilon_0}\ln\rho + \frac{\lambda}{2\pi\varepsilon_0}\ln(2L).\end{aligned}$$

The second term is a constant, which does not contribute to the electric field. We find

$$\vec{\mathbf{E}} = -\nabla\phi \simeq \frac{\lambda}{2\pi\varepsilon_0}\frac{\vec{\rho}}{\rho^2}.$$

Here the vector $\vec{\rho}$ has length ρ and points to the radial direction in the xy-plane. Equivalently, we can view $\vec{\mathbf{E}} = \frac{\lambda}{2\pi\varepsilon_0}\frac{\vec{\rho}}{\rho^2}$ as the electric field produced by the infinite line ($L \to +\infty$) with uniform charge per unit length λ.

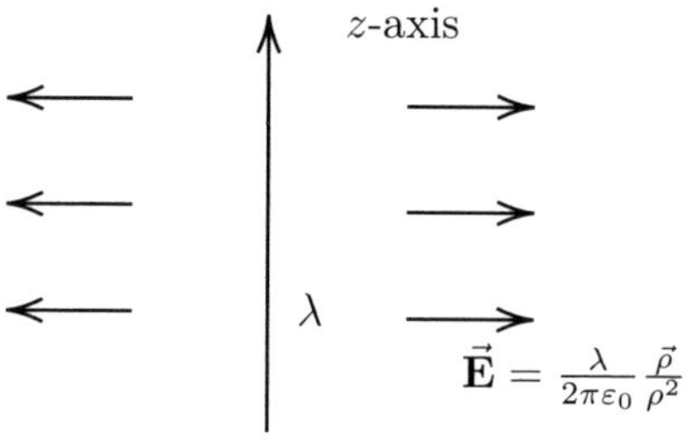

2.2 Interface Condition

2.2.1 Infinite Plane Charge

We consider the electric field produced by surface charge. We first consider the case of an infinite plane located at $z = 0$, carrying uniform charge ξ per unit area. By translation symmetry, we know that the electric field points along the direction of z-axis, only depend on z, and with opposite directions above and below the plane:

$$\vec{\mathbf{E}} = (0, 0, \mathbf{E} = \mathbf{E}(z)), \qquad \mathbf{E}(-z) = -\mathbf{E}(z).$$

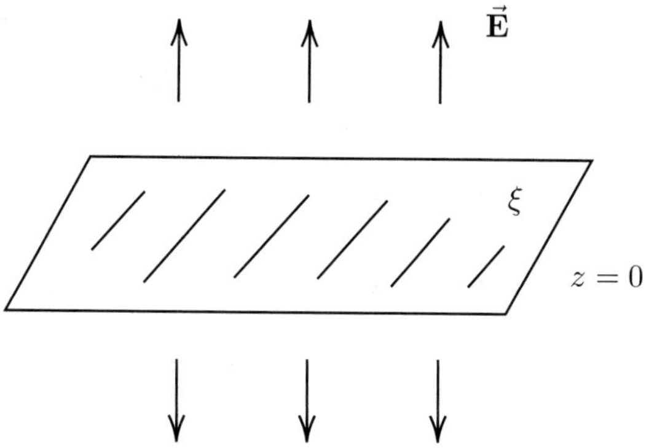

This symmetry is enough to determine $\vec{\mathbf{E}}$. In fact, consider a cylinder Y as below. Let A denote the area of the cap

$$A = \text{Area}(S_\pm) = \text{Area}(S_0).$$

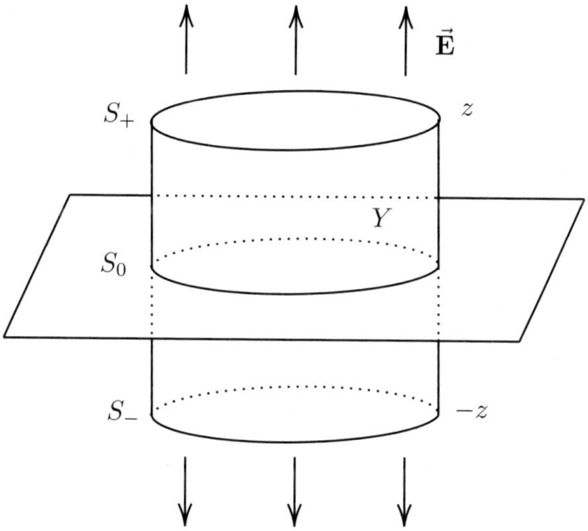

The charge density is

$$\rho = \xi\, \delta(z).$$

Apply Gauss's law to the cylinder

$$\int_{\partial Y} d\vec{\sigma} \cdot \vec{\mathbf{E}} = \frac{1}{\varepsilon_0} \int_Y \rho.$$

This leads to

$$\int_{S_+} d\vec{\sigma} \cdot \vec{\mathbf{E}} - \int_{S_-} d\vec{\sigma} \cdot \vec{\mathbf{E}} = \frac{1}{\varepsilon_0} \int_{S_0} \xi.$$

So

$$\mathbf{E}(z)A - \mathbf{E}(-z)A = \frac{\xi}{\varepsilon_0}A \quad \Longrightarrow \quad \mathbf{E}(z) = \frac{\xi}{2\varepsilon_0} \qquad \text{for} \quad z > 0.$$

Equivalently, we can write

$$\mathbf{E}(z) = \frac{\xi}{2\varepsilon_0}\frac{z}{|z|}.$$

Up to a constant, the scalar potential is

$$\phi = \phi(z) = -\frac{\xi}{2\varepsilon_0}|z|.$$

There is an important observation here: the electric field $\vec{\mathbf{E}}$ is NOT continuous across the charge plane

$$\mathbf{E}(z \to 0^+) - \mathbf{E}(z \to 0^-) = \frac{\xi}{\varepsilon_0}.$$

On the other hand, the scalar potential ϕ is continuous, though its derivative is not. (ϕ can also jump in general, such as in Dipole Layer.)

We discuss some generalization of infinite plane charge. Consider a pair of infinite planes P_+, P_- placed at

$$P_\pm : z = \pm l.$$

$P_\pm$ carries uniform charge $\mp\xi$ per unit area.

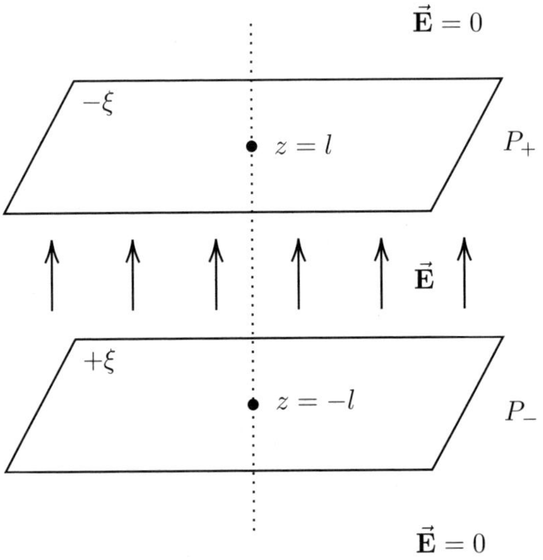

By the result from the infinite plane and the superposition law, we find

$$\vec{\mathbf{E}} = \begin{cases} 0 & |z| > l \\ (0, 0, \xi/\varepsilon_0) & |z| < l \end{cases}$$

As another example, we can consider a bit more realistic model of the infinite plane by an infinite slab of thickness $2l$ with charge density per unit volume ρ, placed at

$$|z| \leq l.$$

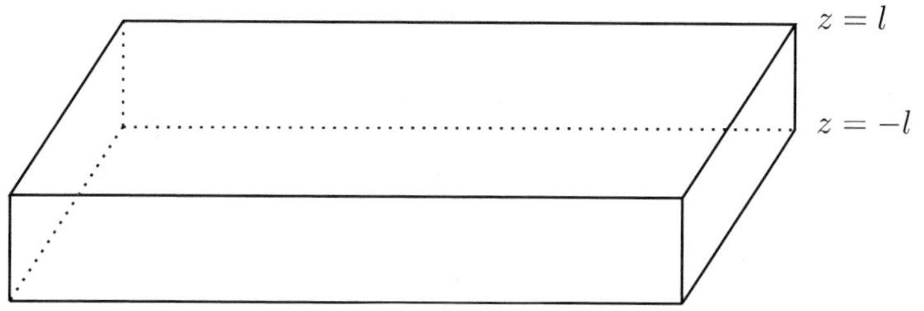

We can think about the effect of each slice of infinite plane and use superposition to sum/integrate them up. Let $\vec{\mathbf{E}} = (0, 0, \mathbf{E}(z))$. When $|z| > L$, the total effect is the same as an infinite plane charge with $\xi = 2l\rho$. Therefore

$$\mathbf{E}(z) = \begin{cases} \dfrac{\rho l}{\varepsilon_0} & z > l \\ -\dfrac{\rho l}{\varepsilon_0} & z < -l \end{cases}.$$

Now we consider a point inside the slab, say at $z = a$ where

$$0 < a < l.$$

The effect of the slab between $a \leq z \leq l$ will cancel that between $-l \leq z \leq -a$.

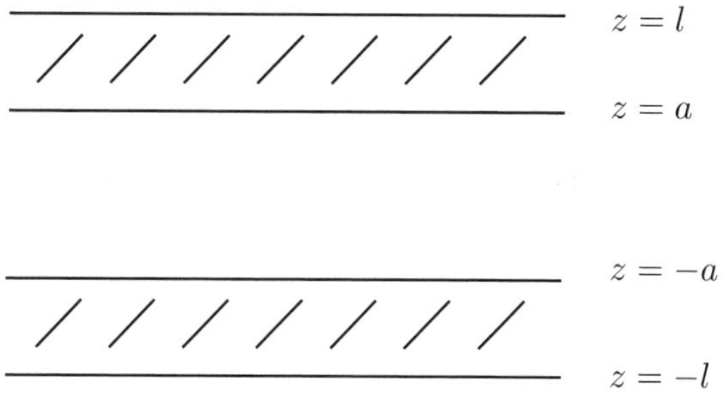

The slab $-a \leq z \leq a$ will contribute

$$\mathbf{E}(a) = \frac{\rho a}{\varepsilon_0}.$$

In summary, we find

$$\mathbf{E}(z) = \begin{cases} \dfrac{\rho l}{\varepsilon_0} & z > l \\ \dfrac{\rho z}{\varepsilon_0} & -l \leq z \leq l \\ -\dfrac{\rho l}{\varepsilon_0} & z < -l \end{cases}.$$

In the limit $l \to 0$, we get back to the case of infinite plane.

2.2.2 Spherical Shell

Consider a spherical shell of radius R with uniform surface charge ξ per area, centered at the origin.

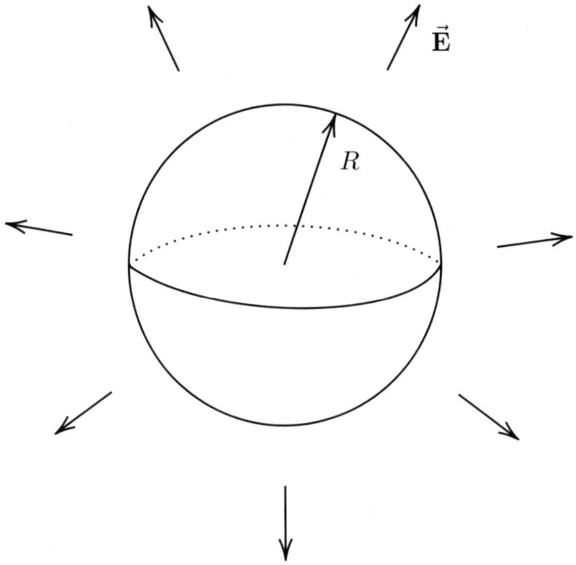

By spherical symmetry,

$$\vec{\mathbf{E}}(\vec{r}) = A(r)\,\vec{r}$$

for some function $A(r)$ of the radius r.

① $\underline{r > R}$. We consider the sphere S_r of radius r which surrounds the charged spherical shell. By Gauss's law

$$\int_{S_r} d\vec{\sigma} \cdot \vec{\mathbf{E}} = Q/\varepsilon_0$$

where $Q = 4\pi R^2 \xi$ is the total charge on the shell. Since

$$\int_{S_r} d\vec{\sigma} \cdot \vec{\mathbf{E}} = A(r) r \,\mathrm{Area}(S_r) = A(r)\, 4\pi r^3 \quad \Longrightarrow \quad A(r) = \frac{Q}{4\pi r^3 \varepsilon_0},$$

i.e.

$$\vec{\mathbf{E}} = \frac{Q}{4\pi\varepsilon_0} \frac{\vec{r}}{r^3}.$$

So effectively it feels the same as a point charge Q.

② $\underline{r < R}$. Again by Gauss's law and spherical symmetry,

$$\int_{S_r} d\vec{\sigma} \cdot \vec{\mathbf{E}} = 0 \quad \Longrightarrow \quad A(r) = 0.$$

So inside the shell

$$\vec{\mathbf{E}} = 0.$$

Note that again we have a jump of $\vec{\mathbf{E}}$ at the surface of the charged shell.

2.2.3 Interface Condition

Now we discuss the relation between the surface charge distribution and the discontinuity of electric field. Consider a surface S with charge density ξ per area. Let $\hat{n}$ denote the unit normal vector on the surface. Let $\vec{\mathbf{E}}_+$ denote the limit of the electric field along the $\hat{n}$ side of the surface, and $\vec{\mathbf{E}}_-$ denote the limit on the other side.

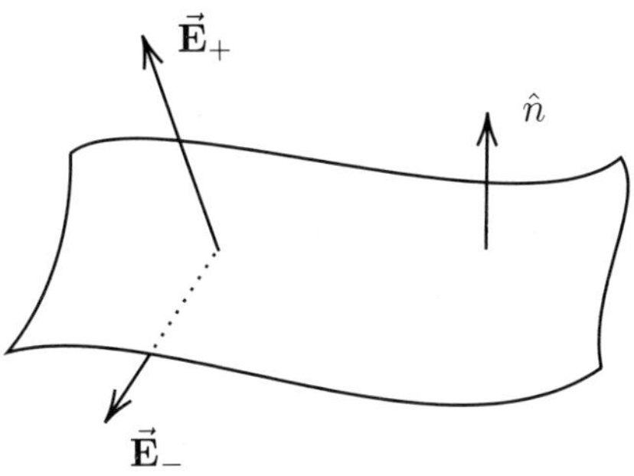

Then the **interface conditions** for electric field along the surface charge distribution is

$$\begin{cases} \hat{n} \cdot (\vec{\mathbf{E}}_+ - \vec{\mathbf{E}}_-) = \xi/\varepsilon_0 & \text{jump along normal direction} \\ \hat{n} \times (\vec{\mathbf{E}}_+ - \vec{\mathbf{E}}_-) = 0 & \text{continuous along tangent direction} \end{cases}$$

We have seen several examples of such phenomenon above. To see how such interface conditions arise, let us first consider a thin cylinder Y across the surface as shown.

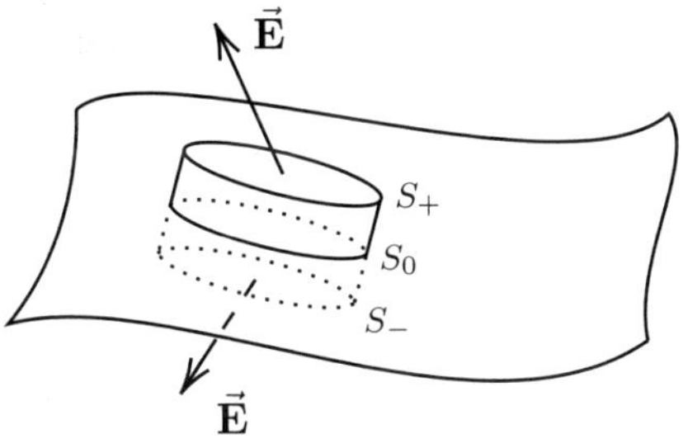

Apply Gauss's law

$$\int_{\partial Y} d\vec{\sigma} \cdot \vec{\mathbf{E}} = \frac{1}{\varepsilon_0} \int_{S_0} \xi.$$

When the cylinder is infinite thin, this leads to

$$\int_{S_0} (\hat{n} \cdot \vec{\mathbf{E}}_+ - \hat{n} \cdot \vec{\mathbf{E}}_-) = \frac{1}{\varepsilon_0} \int_{S_0} \xi.$$

This holds for arbitrary S_0, therefore

$$\hat{n} \cdot (\vec{\mathbf{E}}_+ - \vec{\mathbf{E}}_-) = \frac{\xi}{\varepsilon_0}.$$

To see the continuity of tangent direction, consider the boundary loop C of a rectangle R with a length L parallel to the surface and a short length δ across the surface, as shown below.

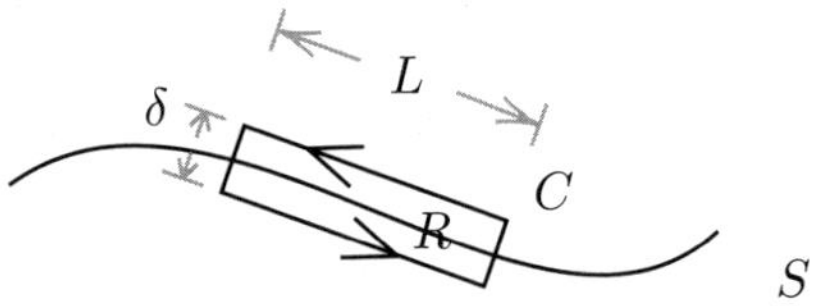

Since $\nabla \times \vec{\mathbf{E}} = 0$, the loop integral

$$\oint_C d\vec{r} \cdot \vec{\mathbf{E}} = \int_R d\vec{\sigma} \cdot (\nabla \times \vec{\mathbf{E}}) = 0.$$

In the limit $\delta \to 0$, this leads to the continuity of the tangent direction of $\vec{\mathbf{E}}$ along S:

$$\hat{n} \times (\vec{\mathbf{E}}_+ - \vec{\mathbf{E}}_-) = 0.$$

There are similar interface conditions for magnetic fields across a surface with current. See Section 2.5.2.

2.2.4 Electric Conductor

A conductor is a material that contains many free electrons that can move in the material but not leave its surface. In "electrostatic" situation, the electrons will move to arrange themselves to produce zero electric field inside the conductor.

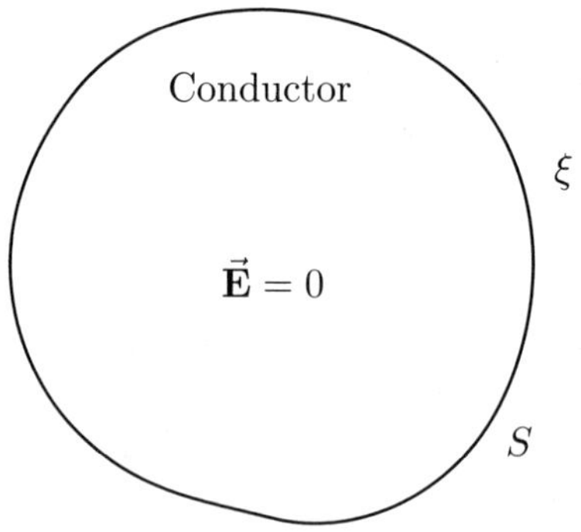

$\vec{\mathbf{E}} = 0$ inside.

On the other hand, its surface S could have a nontrivial charge density ξ per area due to outside situation. For example, we can put a conductor inside two parallel plane charge as shown.

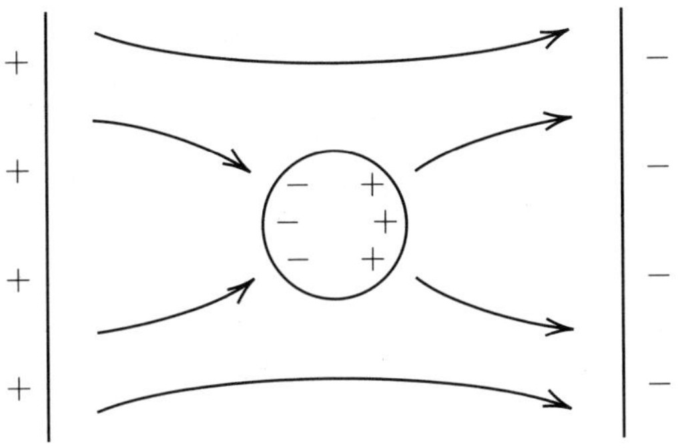

Let $\hat{n}$ denote the unit normal vector on S that points outward.

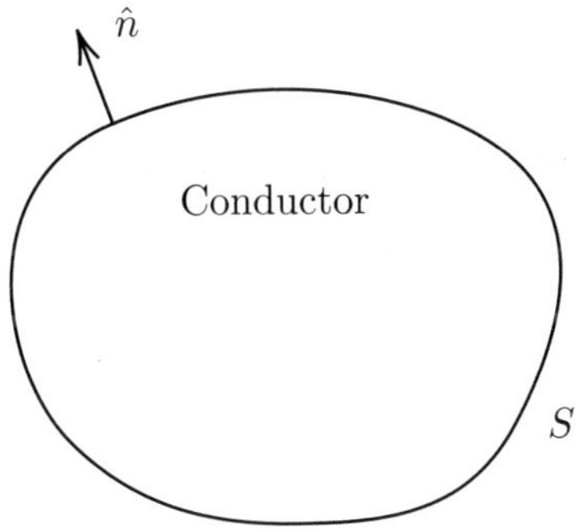

Then we have inside the conductor

$$\vec{\mathbf{E}}_- = 0.$$

The continuity of tangent direction implies that $\vec{\mathbf{E}}_+$ is normal to S. This is easy to understand: if there is a nontrivial tangent direction, then it will further move electrons on the surface. Now from the interface condition of the normal direction, we find on the outside surface of the conductor

$$\vec{\mathbf{E}}_+ = \frac{\xi}{\varepsilon_0}\hat{n}.$$

Since $\vec{\mathbf{E}} = 0$ inside the conductor, the scalar potential ϕ is a constant inside. In the outside, we have on the surface of the conductor

$$\partial_{\hat{n}}\phi|_S = -\frac{\xi}{\varepsilon_0}.$$

2.3 Boundary Value Problem

In electrostatics, we are led to consider the scalar potential ϕ satisfying the Poisson's equation

$$\nabla^2\phi = -\rho/\varepsilon_0$$

subject to certain boundary condition in a region V under consideration. There are usually two types of boundary conditions on $S = \partial V$.
① Dirichlet boundary condition: specify the value of ϕ on the boundary

$$\phi|_S = f.$$

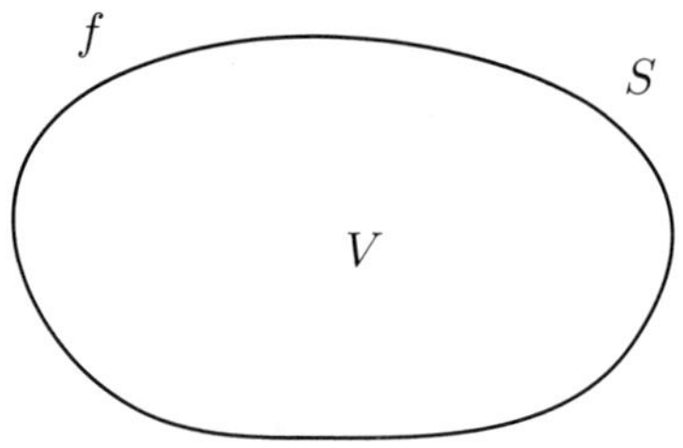

② Neumann boundary condition: specify the value of the normal derivative on the boundary

$$\partial_{\hat{n}}\phi|_S = g.$$

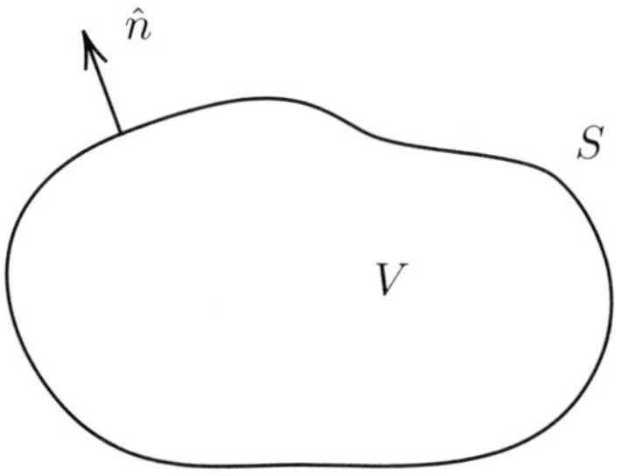

2.3.1 Uniqueness

We first show the uniqueness of the solution of the Poisson's equation

$$\nabla^2\phi = -\rho/\varepsilon_0$$

inside a finite region V subject to either Dirichlet or Neumann boundary conditions.

Suppose we have two solutions, say ϕ_1 and ϕ_2. Let $\gamma = \phi_2 - \phi_1$. Then γ satisfies the Laplace's equation

$$\nabla^2\gamma = 0$$

and either $\gamma|_S = 0$ (Dirichlet case) or $\partial_{\hat{n}}\gamma|_S = 0$ (Neumann case). Consider the volume integral (we omit the volume measure d^3r on V and also the surface measure on S for simplicity)

$$\begin{aligned}\int_V |\nabla\gamma|^2 &= \int_V \nabla\cdot(\gamma\nabla\gamma) - \gamma\nabla^2\gamma \\ &= \int_S d\vec{\sigma}\cdot(\gamma\nabla\gamma) - \int_V \gamma\nabla^2\gamma \\ &= \int_S \gamma\partial_{\hat{n}}\gamma - \int_V \gamma\nabla^2\gamma.\end{aligned}$$

Either Dirichlet or Neumann condition implies $\gamma\partial_{\hat{n}}\gamma = 0$. Together with the Laplace equation $\nabla^2\gamma = 0$, it follows that

$$\int_V |\nabla\gamma|^2 = 0.$$

This implies $\nabla\gamma = 0$, hence γ is a constant. For Dirichlet boundary condition, $\gamma = 0$ and the solution is unique. For Neumann boundary condition, the solution is unique up to an additive constant.

2.3.2 Green's Function

The solution to the Poisson's equation in a finite region V with either Dirichlet or Neumann boundary condition can be obtained by means of Green's function $G(\vec{r},\vec{r}\,')$, which satisfies

$$\nabla^{2\prime}G(\vec{r},\vec{r}\,') = -4\pi\delta(\vec{r}-\vec{r}\,'), \qquad \vec{r},\vec{r}\,' \in V.$$

Here $\nabla^{2\prime}$ is the Laplacian with respect to $\vec{r}\,'$. Such a Green's function can be written as

$$G(\vec{r},\vec{r}\,') = \frac{1}{|\vec{r}-\vec{r}\,'|} + F(\vec{r},\vec{r}\,')$$

where F satisfies the Laplace's equation

$$\nabla^{2\prime}F(\vec{r},\vec{r}\,') = 0 \qquad \text{inside } V.$$

Consider $\phi(\vec{r})$ which satisfies the Poisson's equation

$$\nabla^2\phi = -\rho/\varepsilon_0 \qquad \text{inside } V.$$

Then for $\vec{r}$ inside V, we have

$$\begin{aligned}\phi(\vec{r}) &= \int_V d^3r'\,\phi(\vec{r}\,')\delta(\vec{r}-\vec{r}\,') = -\frac{1}{4\pi}\int_V d^3r'\,\phi(\vec{r}\,')\nabla^{2\prime}G(\vec{r},\vec{r}\,') \\ &\overset{\text{integration}}{\underset{\text{by part}}{=\!=\!=}} \frac{1}{4\pi}\int_V d^3r'\,\nabla'\phi(\vec{r}\,')\cdot\nabla'G(\vec{r},\vec{r}\,') - \frac{1}{4\pi}\int_S \phi(\vec{r}\,')\partial'_{\hat{n}}G(\vec{r},\vec{r}\,') \\ &= -\frac{1}{4\pi}\int_V d^3r'\,\nabla^{2\prime}\phi(\vec{r}\,')G(\vec{r},\vec{r}\,') + \frac{1}{4\pi}\int_S \left[\partial'_{\hat{n}}\phi(\vec{r}\,')G(\vec{r},\vec{r}\,') - \phi(\vec{r}\,')\partial'_{\hat{n}}G(\vec{r},\vec{r}\,')\right] \\ &= \frac{1}{4\pi\varepsilon_0}\int_V d^3r'\,\rho(\vec{r}\,')G(\vec{r},\vec{r}\,') + \frac{1}{4\pi}\int_S \left[\partial'_{\hat{n}}\phi(\vec{r}\,')G(\vec{r},\vec{r}\,') - \phi(\vec{r}\,')\partial'_{\hat{n}}G(\vec{r},\vec{r}\,')\right].\end{aligned}$$

There are two types of Green's function depending on the type of boundary conditions.
① Dirichlet Green's function $G_D(\vec{r},\vec{r}\,')$.

$$\begin{cases} \nabla^{2\prime} G_D(\vec{r},\vec{r}\,') = -4\pi\delta(\vec{r}-\vec{r}\,') \\ G_D(\vec{r},\vec{r}\,') = 0 \qquad \text{for } \vec{r}\,' \in S \end{cases}$$

Using $G_D(\vec{r},\vec{r}\,')$, we find

$$\phi(\vec{r}) = \frac{1}{4\pi\varepsilon_0}\int_V \rho(\vec{r}\,')G_D(\vec{r},\vec{r}\,') - \frac{1}{4\pi}\int_S \phi(\vec{r}\,')\partial_{\hat{n}}{}' G_D(\vec{r},\vec{r}\,')$$

which gives an explicit formula of ϕ in terms of its boundary value $\phi|_S$. Thus this formula is suitable for Dirichlet boundary value problem.

Remark 2.3.1. $G_D(\vec{r},\vec{r}\,')$ has the symmetry property in general

$$G_D(\vec{r},\vec{r}\,') = G_D(\vec{r}\,',\vec{r}).$$

② Neumann Green's function $G_N(\vec{r},\vec{r}\,')$.

$$\begin{cases} \nabla^{2\prime} G_N(\vec{r},\vec{r}\,') = -4\pi\delta(\vec{r}-\vec{r}\,') \\ \partial_{\hat{n}}{}' G_N(\vec{r},\vec{r}\,') = -4\pi/A \qquad \text{for } \vec{r}\,' \in S \end{cases}$$

Here $A = \text{Area}(S)$ is the total area of the boundary surface S. Such choice of constant is to be consistent with the equation $\nabla^{2\prime} G_N(\vec{r},\vec{r}\,') = -4\pi\delta(\vec{r}-\vec{r}\,')$. In fact,

$$-4\pi = \int_V d^3r'\,\nabla^{2\prime} G_N(\vec{r},\vec{r}\,') = \int_S \partial_{\hat{n}}{}' G_N(\vec{r},\vec{r}\,') = \int_S (-4\pi)/A = -4\pi\frac{\int_S 1}{A}$$

$$\Longrightarrow \qquad A = \int_S 1 = \text{Area}(S).$$

Using $G_N(\vec{r},\vec{r}\,')$, we find

$$\begin{aligned} \phi(\vec{r}) &= \frac{1}{4\pi\varepsilon_0}\int_V d^3r'\,\rho(\vec{r}\,')G_N(\vec{r},\vec{r}\,') + \frac{1}{4\pi}\int_S \partial'_{\hat{n}}\phi(\vec{r}\,')G_N(\vec{r},\vec{r}\,') \\ &\quad - \frac{1}{4\pi}\int_S \phi(\vec{r}\,')\partial_{\hat{n}}{}' G_N(\vec{r},\vec{r}\,') \\ &= \frac{1}{A}\int_S \phi(\vec{r}\,') + \frac{1}{4\pi\varepsilon_0}\int_V d^3r'\,\rho(\vec{r}\,')G_N(\vec{r},\vec{r}\,') + \frac{1}{4\pi}\int_S \partial'_{\hat{n}}\phi(\vec{r}\,')G_N(\vec{r},\vec{r}\,'). \end{aligned}$$

which gives an explicit formula of ϕ in terms of its boundary normal derivative $\partial_{\hat{n}}\phi|_S$ and the boundary integral $\int_S \phi(\vec{r}\,')$. Thus this formula is suitable for Neumann boundary value problem.

We can also interpret Green's function $G(\vec{r}, \vec{r}\,')$ as an electrostatic problem as follows. Let us consider the following potential function $\phi_{\vec{r}}$

$$\phi_{\vec{r}}(\vec{r}\,') := \frac{1}{4\pi\varepsilon_0} G(\vec{r}, \vec{r}\,').$$

Then it satisfies the equation

$$\nabla^{2\prime} \phi_{\vec{r}}(\vec{r}\,') = -\frac{1}{\varepsilon_0}\delta(\vec{r} - \vec{r}\,')$$

which represents a potential for a unit point charge located at $\vec{r}$ inside V. Let us write

$$\phi_{\vec{r}}(\vec{r}\,') = \frac{1}{4\pi\varepsilon_0}\frac{1}{|\vec{r} - \vec{r}\,'|} + \xi_{\vec{r}}(\vec{r}\,')$$

for some $\xi_{\vec{r}}(\vec{r}\,')$. Then $\xi_{\vec{r}}(\vec{r}\,')$ satisfies the equation

$$\nabla^{2\prime} \xi_{\vec{r}}(\vec{r}\,') = 0.$$

It represents the potential due to certain external charge distribution outside V chosen in such a way that the combined potential with the internal point charge at $\vec{r}$ has the specific boundary condition on ∂V.

Remark 2.3.2. **Green's function on $\mathbb{R}^n$.** A strategy to solve the Poisson's equation on $\mathbb{R}^n$ is to find a fundamental solution of

$$\nabla^2 \Phi(\vec{r}) = -\delta(\vec{r}).$$

Suppose $\Phi = \Phi(r)$ depends only on $r = |\vec{r}|$. Then

$$\nabla^2\Phi(r) = \Phi''(r) + \frac{n-1}{r}\Phi'(r) = 0 \qquad \Longleftrightarrow \qquad (\ln(\Phi'(r)))' = -\frac{n-1}{r}.$$

The solution of $\nabla^2\Phi(r) = 0$ on $\mathbb{R}^n - \{0\}$ is

$$\Phi(r) = \begin{cases} c_1 \ln r + c_2 & n = 2 \\ c_1 r^{-n+2} + c_2 & n \geq 3 \end{cases}$$

where c_i's are constants. We choose appropriate constants such that

$$\Phi(r) = \begin{cases} -\dfrac{1}{2\pi}\ln r & n = 2 \\ \dfrac{1}{n(n-2)\alpha(n) r^{n-2}} & n \geq 3 \end{cases}$$

Here $\alpha(n) = \dfrac{\pi^{n/2}}{\Gamma(1 + n/2)}$ is the volume of the n-dimensional unit ball, and

$$\Gamma(1 + n/2) = \begin{cases} (n/2)! & n \text{ even} \\ n!!2^{-(n+1)/2}\sqrt{\pi} & n \text{ odd} \end{cases}$$

Then we show $\nabla^2\Phi = -\delta(\vec{r})$ in the sense of distribution. This means for each smooth function $g : \mathbb{R}^n \to \mathbb{R}$ with compact support, we have

$$\int_{\mathbb{R}^n} \Phi \cdot \nabla^2 g = -g(0).$$

Indeed,

$$\begin{aligned}
\int_{\mathbb{R}^n} \Phi \cdot \nabla^2 g &= \int_{B(0,\varepsilon)} \Phi \cdot \nabla^2 g + \int_{B(0,\varepsilon)^c} \Phi \cdot \nabla^2 g \\
&\overset{\text{Integration}}{\underset{\text{by part}}{=\!=\!=}} \underbrace{\int_{B(0,\varepsilon)} \Phi \cdot \nabla^2 g}_{\leq c||\nabla g||_{L^\infty} \int_{B(0,\varepsilon)} r\Phi} + \underbrace{\int_{B(0,\varepsilon)^c} \nabla^2\Phi \cdot g}_{=0} - \underbrace{\int_{\partial(B(0,\varepsilon)^c)} \partial_{\vec{n}}\Phi \cdot g}_{=\frac{\int_{\partial(B(0,\varepsilon)^c)} g}{\mathrm{Vol}(\partial(B(0,\varepsilon)^c))}} \\
&\quad + \underbrace{\int_{\partial(B(0,\varepsilon)^c)} \Phi \cdot \partial_{\vec{n}} g}_{\leq ||\partial_{\vec{n}} g||_{L^\infty} \int_{\partial(B(0,\varepsilon)^c)} \Phi} \\
&\overset{\varepsilon \to 0}{=\!=\!=} 0 + 0 - g(0) + 0 = -g(0).
\end{aligned}$$

Here $B(0,\varepsilon)$ is the ball of radius ε centered at 0, and $B(0,\varepsilon)^c = \mathbb{R}^n - B(0,\varepsilon)$. It follows that

$$\phi(\vec{r}) = \frac{1}{\varepsilon_0} \int_{\mathbb{R}^n} d\vec{r}\,'\Phi(\vec{r}\,')\rho(\vec{r} - \vec{r}\,')$$

is a solution to the Poisson's equation

$$\nabla^2\phi = -\rho/\varepsilon_0.$$

2.3.3 Method of Image

Let us first consider an example where the region is

$$V = \{z \geq 0\}.$$

We would like to figure out $G_D(\vec{r}, \vec{r}\,')$. Let us put a point charge q at

$$\vec{r}_+ = (x_0, y_0, z_0 > 0)$$

and assume that $S = \{z = 0\}$ is a conducting plane (so the scalar potential is constant on S).

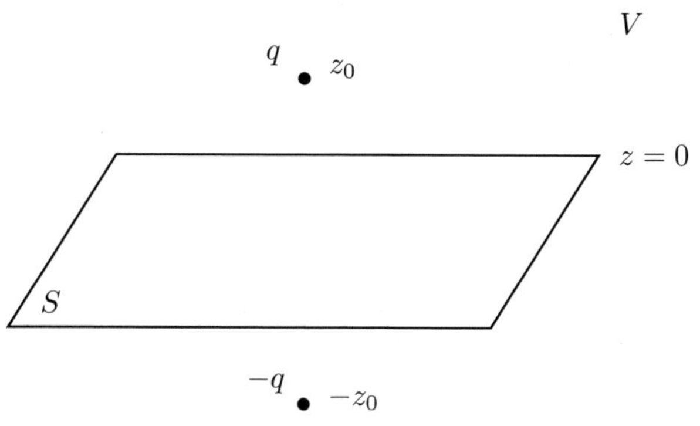

We look for a scalar potential $\phi_{\vec{r}_+}$ such that

$$\begin{cases} \nabla^2 \phi_{\vec{r}_+}(\vec{r}) = -\dfrac{\delta(\vec{r}-\vec{r}_+)}{\varepsilon_0} \\ \phi_{\vec{r}_+}(\vec{r}) = 0 \qquad \text{when} \quad \vec{r} \in S \end{cases}$$

This can be solved by replacing the conductor by a fictitious "image" point charge $-q$ at the point $\vec{r}_- = (x_0, y_0, -z_0)$. Then

$$\phi_{\vec{r}_+}(\vec{r}) = \frac{q}{4\pi\varepsilon_0 |\vec{r}-\vec{r}_+|} - \frac{q}{4\pi\varepsilon_0 |\vec{r}-\vec{r}_-|}$$

satisfies the above Dirichlet boundary value problem in V, hence is the unique solution. As a consequence, the Dirichlet Green's function is

$$G_D(\vec{r}, \vec{r}\,') = \frac{1}{|\vec{r}-\vec{r}\,'|} - \frac{1}{|\vec{r}-\sigma(\vec{r}\,')|}$$

where $\sigma : (x, y, z) \mapsto (x, y, -z)$ is the mirror transformation. Using $|\vec{r}-\sigma(\vec{r}\,')| = |\sigma(\vec{r}) - \vec{r}\,'|$, we see explicitly the symmetry property

$$G_D(\vec{r}, \vec{r}\,') = G_D(\vec{r}\,', \vec{r}).$$

Here we can think about the mirror charge $\sigma(\vec{r}\,')$ as an external charge outside V (the $F(\vec{r}, \vec{r}\,')$ term as discussed in Section 2.3.2).

As another example, we consider a point charge q outside a conducting sphere of radius R. We set the potential on the conducting sphere to be zero. Finding the potential of this problem is the same as describing the Dirichlet Green's function on $V = \{|\vec{r}| \geq R\}$, $S = \{|\vec{r}| = R\}$.

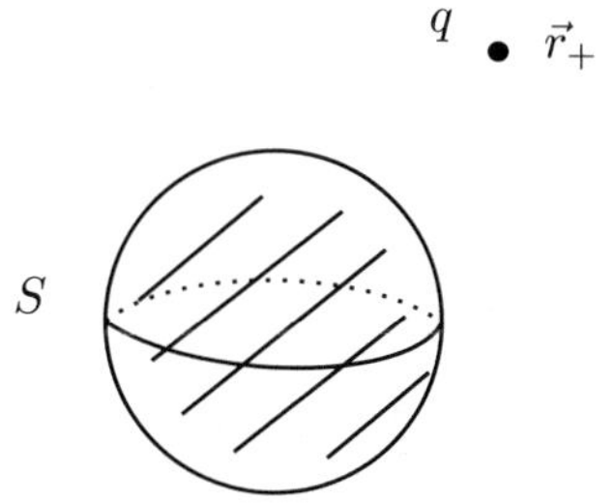

By the same image method, we can guess the result by

$$\phi_{\vec{r}_+}(\vec{r}) = \frac{q}{4\pi\varepsilon_0}\left(\frac{1}{|\vec{r}-\vec{r}_+|} - \frac{R/|\vec{r}_+|}{|\vec{r}-\sigma_R(\vec{r}_+)|}\right).$$

Here $\sigma_R : \vec{r} \mapsto \frac{R^2}{|\vec{r}|^2}\vec{r}$ is the mirror image of $\vec{r}$ with respect to the sphere of radius R.

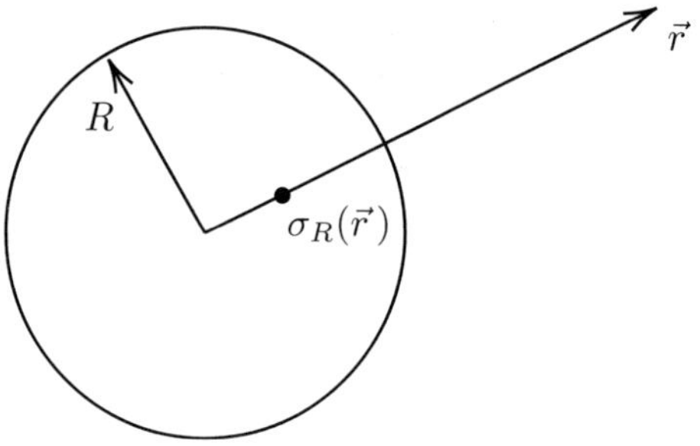

It is clear that

$$\nabla^2\phi_{\vec{r}_+}(\vec{r}) = -\frac{q}{4\pi\varepsilon_0}\delta(\vec{r}-\vec{r}_+) \qquad \text{inside } V.$$

On the other hand, let $\hat{n} = \vec{r}/|\vec{r}|$ and $\hat{n}_+ = \vec{r}_+/|\vec{r}_+|$ denote the unit direction of $\vec{r}$, $\vec{r}_+$ and let $r = |\vec{r}|$, $r_+ = |\vec{r}_+|$. Then

$$\phi_{\vec{r}_+}(\vec{r}) = \frac{q}{4\pi\varepsilon_0}\left(\frac{1}{|\vec{r}-\vec{r}_+|} - \frac{R}{|rr_+\hat{n} - R^2\hat{n}_+|}\right).$$

It is clear from this expression that

$$\phi_{\vec{r}_+}(\vec{r}) = \phi_{\vec{r}}(\vec{r}_+).$$

It follows that

$$\phi_{\vec{r}_+}(\vec{r})|_{\vec{r}\in S} = \phi_{\vec{r}}(\vec{r}_+)|_{\vec{r}\in S} = 0.$$

Thus the Dirichlet Green's function on V is

$$G_D(\vec{r},\vec{r}\,') = \frac{1}{|\vec{r}-\vec{r}\,'|} - \frac{R/|\vec{r}\,'|}{|\vec{r}-\sigma_R(\vec{r}\,')|}$$

which is indeed symmetric: $G_D(\vec{r},\vec{r}\,') = G_D(\vec{r}\,',\vec{r})$.

2.4 Electric Dipole and Dielectric Polarisation

2.4.1 Electric Multipole Expansion

Assume we have a charge distribution ρ in a finite region V. It generates a potential via

$$\phi(\vec{r}) = \frac{1}{4\pi\varepsilon_0}\int_V d^3r'\,\frac{\rho(\vec{r}\,')}{|\vec{r}-\vec{r}\,'|}.$$

We consider an approximation to $\phi(\vec{r})$ when $\vec{r}$ lies far from the region V. This is so-called **electric multipole expansion**. Here we assume the region V is inside a sphere of radius R, then by "far", we mean $r >> R$. This implies $r' << r$ in the domain of integration $\int_V d^3r'$.

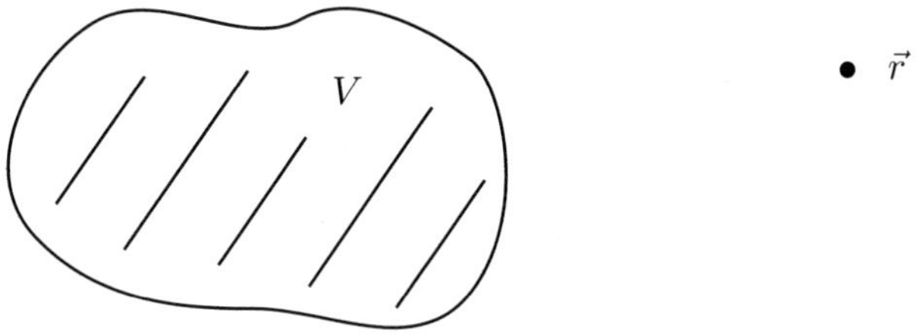

In the region $r' << r$, we have valid Taylor series expansion

$$\frac{1}{|\vec{r}-\vec{r}\,'|} = \frac{1}{r} - \vec{r}\,' \cdot \nabla\left(\frac{1}{r}\right) + \cdots .$$

Let us write it in the following form

$$\frac{1}{|\vec{r}-\vec{r}\,'|} = \frac{1}{\sqrt{r^2 - 2\vec{r}\cdot\vec{r}\,' + (r')^2}} = \frac{1}{r\sqrt{1 - 2\,(\hat{r}\cdot\hat{r}\,')\,\frac{r'}{r} + \left(\frac{r'}{r}\right)^2}}$$

where $\hat{r} = \frac{\vec{r}}{r}$ is the unit vector in the direction of $\vec{r}$.

Recall the series expansion

$$\frac{1}{\sqrt{1-2xt+t^2}} = \sum_{l=0}^{\infty} t^l P_l(x) \qquad \text{for } |x| \le 1,\ 0 < t < 1.$$

Here $P_l(x)$'s are the so-called **Legendre polynomials**. Explicitly, they are given by

$$P_l(x) = \frac{1}{2^l l!}\frac{d^l}{dx^l}(x^2-1)^l.$$

The first a few terms are

$$P_0(x) = 1, \qquad P_1(x) = x,$$
$$P_2(x) = \frac{1}{2}(3x^2-1), \qquad P_3(x) = \frac{1}{2}(5x^3-3x),$$
$$\ldots$$

These Legendre polynomials are orthogonal in the sense

$$\int_{-1}^{1} dx\, P_l(x) P_m(x) = \frac{2}{2l+1}\delta_{lm}$$

and complete in the sense

$$\sum_{l=0}^{\infty}\left(l+\frac{1}{2}\right)P_l(x)P_l(x')=\delta(x-x').$$

In terms of Legendre polynomials,

$$\frac{1}{|\vec{r}-\vec{r}\,'|}=\frac{1}{r}\sum_{l=0}^{\infty}\left(\frac{r'}{r}\right)^l P_l(\hat{r}\cdot\hat{r}\,'),$$

from which we find the multipole expansion

$$\begin{aligned}
\phi(\vec{r}) &= \frac{1}{4\pi\varepsilon_0 r}\int_V d^3r'\sum_{l=0}^{\infty}\left(\frac{r'}{r}\right)^l P_l(\hat{r}\cdot\hat{r}\,')\rho(\vec{r}\,') \\
&= \frac{1}{4\pi\varepsilon_0 r}\int_V d^3r'\left[\rho(\vec{r}\,')+\frac{\vec{r}\cdot\vec{r}\,'}{r^2}\rho(\vec{r}\,')+\left(\frac{r'}{r}\right)^2\frac{1}{2}\left(3\left(\frac{\vec{r}\cdot\vec{r}\,'}{rr'}\right)^2-1\right)\rho(\vec{r}\,')+\cdots\right] \\
&= \frac{1}{4\pi\varepsilon_0}\left[\frac{Q}{r}+\frac{\vec{P}\cdot\vec{r}}{r^3}+\sum_{i,j}Q_{ij}\frac{3r_ir_j-r^2\delta_{ij}}{r^5}+\cdots\right]
\end{aligned}$$

where

$$Q=\int_V d^3r'\,\rho(\vec{r}\,')$$

is the total electric charge, and

$$\vec{P}=\int_V d^3r'\,\rho(\vec{r}\,')\vec{r}\,'$$

is the **electric dipole moment**, and

$$Q_{ij}=\frac{1}{2}\int_V d^3r'\,\rho(\vec{r}\,')r_i'r_j'$$

is the electric quadrupole where we write the vector $\vec{r}\,'$ as $\vec{r}\,'=(r_1',r_2',r_3')$.

For electrically neutral object ($Q=0$), the second term describes the leading electrostatic potential at far distance

$$\phi(\vec{r})\simeq\frac{1}{4\pi\varepsilon_0}\frac{\vec{P}\cdot\vec{r}}{r^3},\qquad r>>0.$$

Then the leading approximation of the electric field is

$$\mathbf{\vec{E}}=-\nabla\phi(\vec{r})\simeq\frac{1}{4\pi\varepsilon_0}\frac{3(\hat{r}\cdot\vec{P})\hat{r}-\vec{P}}{r^3},\qquad r>>0.$$

2.4.2 The Electric Dipole

Consider two point charges $+q$ and $-q$ at a distance d apart. The total charge is zero. The dipole moment is

$$\vec{P} = q\vec{d}.$$

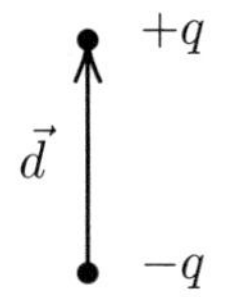

The "**electric dipole**" is the limit $d \to 0$ and $q \to +\infty$ such that $\vec{P}$ is fixed. In this limit, the potential is

$$\phi(\vec{r}) = \frac{1}{4\pi\varepsilon_0}\frac{\vec{P}\cdot\vec{r}}{r^3}$$

where higher terms in the multipole expansion vanish in the limit. The electric field is

$$\vec{\mathbf{E}}(\vec{r}) = \frac{1}{4\pi\varepsilon_0}\frac{3(\hat{r}\cdot\vec{P})\hat{r} - \vec{P}}{r^3}.$$

This allows us to draw the electric field of the dipole as

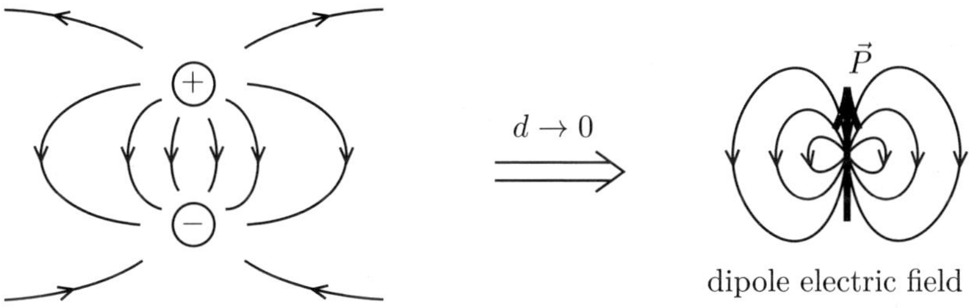

dipole electric field

2.4.3 Dielectric Matter

A **dielectric** is an electrical insulator that can be polarized by an applied electric field. In contrast to conductors, dielectrics do not have charges that are free to move around, and are typically neutral. Instead, an applied external electric field $\vec{\mathbf{E}}_{\text{ext}}$ will cause dielectric polarisation that creates an internal electric field $\vec{\mathbf{E}}_{\text{int}}$ to reduce the overall field within the dielectric. Unlike the conductor, the total macroscopic field is non-zero both inside and outside the dielectric:

$$\vec{\mathbf{E}}_{\text{tot}} = \vec{\mathbf{E}}_{\text{ext}} + \vec{\mathbf{E}}_{\text{int}} \neq 0 \qquad \text{inside dielectric.}$$

Let us consider a simple model of a lattice of neutral atoms.

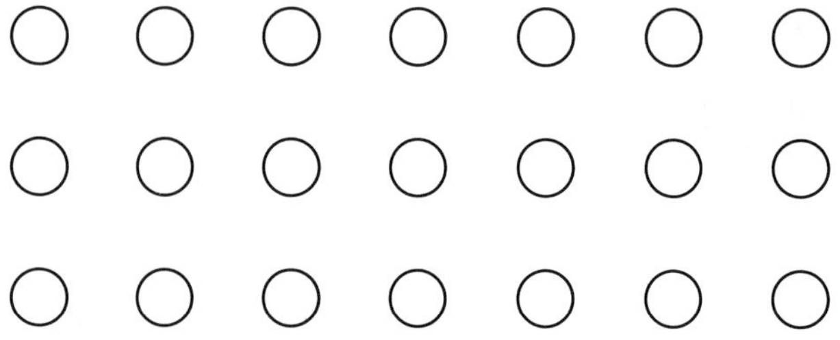

Each nucleus has charge $+q$ at the center surrounded by a spherical cloud of electrons of charge $-q$ such that the total effect is neutral. Let us now apply an external electric field, and the atoms get polarized

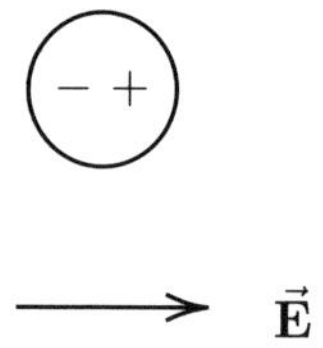

with a dipole moment. This can be approximated by electric dipoles. The **polarisation** $\vec{P}(\vec{r})$ is the dipole moment density per unit volume. The potential arising from the polarisation is given by integrating the dipole potential

$$\begin{aligned}\phi(\vec{r}) &= \frac{1}{4\pi\varepsilon_0}\int_V d^3r'\,\frac{\vec{P}(\vec{r}\,')\cdot(\vec{r}-\vec{r}\,')}{|\vec{r}-\vec{r}\,'|^3} = \frac{1}{4\pi\varepsilon_0}\int_V d^3r'\,\vec{P}(\vec{r}\,')\cdot\nabla'\left(\frac{1}{|\vec{r}-\vec{r}\,'|}\right)\\ &= \frac{1}{4\pi\varepsilon_0}\int_{\partial V} d\vec{\sigma}\cdot\frac{\vec{P}(\vec{r}\,')}{|\vec{r}-\vec{r}\,'|} - \frac{1}{4\pi\varepsilon_0}\int_V d^3r'\,\frac{\nabla'\cdot\vec{P}(\vec{r}\,')}{|\vec{r}-\vec{r}\,'|}.\end{aligned}$$

The first surface integral is the potential from the surface charge density

$$\vec{P}\cdot\hat{n}$$

where $\hat{n}$ is the unit normal vector on ∂V.

The second term is the potential from the charge distribution inside the matter

$$\rho_{\text{bound}}(\vec{r}) = -\nabla\cdot\vec{P}(\vec{r}).$$

This is called the **bound charge** due to polarisation. Assume there are also some charges inside the matter that are free to move, which do not arise form polarisation. Let us call this extra charge ρ_{free}. Then total potential consists of that arising from the polarisation and that produced from the free charges. Thus the total electric field is

$$\nabla\cdot\vec{\mathbf{E}} = \frac{1}{\varepsilon_0}\rho = \frac{1}{\varepsilon_0}(\rho_{\text{free}}+\rho_{\text{bound}}) = \frac{1}{\varepsilon_0}(\rho_{\text{free}}-\nabla\cdot\vec{P}).$$

Define the **electric displacement**

$$\vec{D} = \varepsilon_0\vec{\mathbf{E}} + \vec{P}.$$

Then it obeys

$$\nabla \cdot \vec{D} = \rho_{\text{free}}.$$

It turns out that $\vec{P}$ is proportional to $\vec{\mathbf{E}}$ for most materials, which are called **linear dielectrics**. Let us write

$$\vec{P} = \varepsilon_0 \chi_e \vec{\mathbf{E}}$$

where $\chi_e > 0$ is called the **electric susceptibility**. Then

$$\vec{D} = \varepsilon_0 (1 + \chi_e) \vec{\mathbf{E}} = \varepsilon \vec{\mathbf{E}}$$

where $\varepsilon > \varepsilon_0$ is called the **permittivity** of the material. $\varepsilon_r = \varepsilon / \varepsilon_0 = 1 + \chi_e$ is called the relative permittivity. Then

$$\nabla \cdot \vec{\mathbf{E}} = \frac{\rho_{\text{free}}}{\varepsilon}$$

takes the same form as that in the vacuum. The polarisation has simply the effect of replacing ε_0 by ε.

Example 2.4.1. Consider a dielectric sphere of radius R, with a point charge Q at the center.

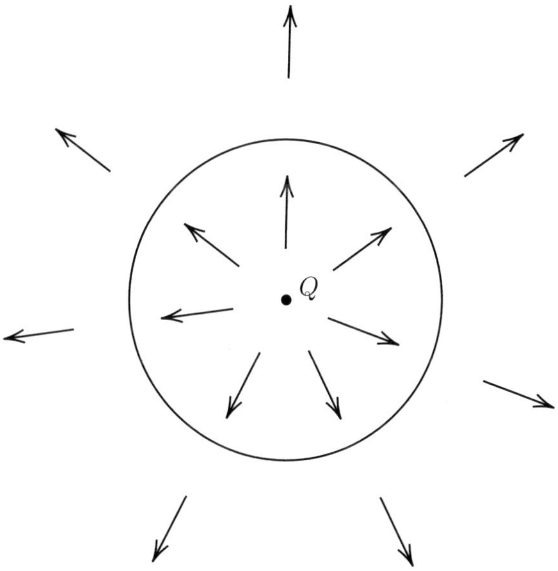

Then the free charge is

$$\rho_{\text{free}} = Q\delta(\vec{r}).$$

Gauss's law inside the dielectric

$$\nabla \cdot \vec{D} = \rho_{\text{free}}$$

implies (as we have seen before)

$$\vec{D} = \frac{Q}{4\pi r^3} \vec{r} \qquad (r < R)$$

The electric field is then

$$\vec{\mathbf{E}} = \vec{D}/\varepsilon = \frac{1}{\varepsilon_r}\frac{Q}{4\pi\varepsilon_0 r^3}\vec{r} \qquad (r < R)$$

This behaves like a charge Q/ε_r placed at the origin. This says that the bound charge gathers at the origin, screening the original charge.

Outside the matter, we have

$$\vec{\mathbf{E}} = \frac{Q}{4\pi\varepsilon_0 r^3}\vec{r} \qquad (r > R)$$

which does not depend on the polarization. On the surface, we have

$$\vec{\mathbf{E}}_+ = \frac{Q}{4\pi\varepsilon_0 R^2}\hat{n}, \qquad \vec{\mathbf{E}}_- = \frac{Q}{4\pi\varepsilon R^2}\hat{n}.$$

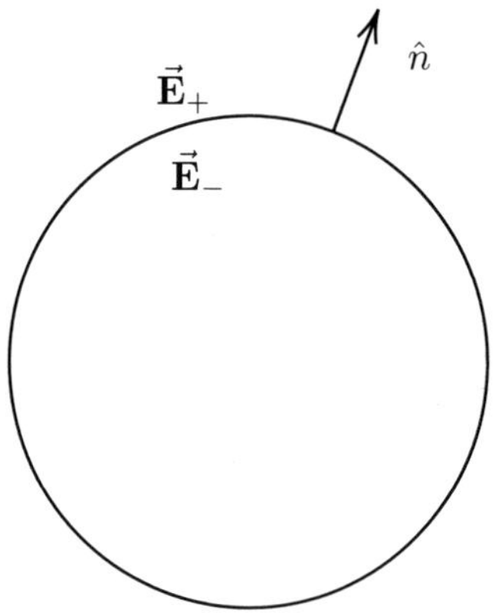

From the interface condition, we have a surface charge density

$$\vec{\mathbf{E}}_+ - \vec{\mathbf{E}}_- = \frac{\xi}{\varepsilon_0}\hat{n},$$

$$\xi = \frac{Q}{4\pi R^2}\left(1 - \frac{1}{\varepsilon_r}\right).$$

The total surface charge is therefore $Q - Q/\varepsilon_r$. This is precisely the opposite of the bound charge at the origin, as expected.

2.5 Magnetic Field and Vector Potential

We move on to study magnetostatics. This is the case when we have a steady current $\vec{\mathbf{j}}$. As we have seen in the discussion of Maxwell's equation:

charge	$\Longrightarrow$	electric field
current	$\Longrightarrow$	magnetic field

Recall the conservation law of charge

$$\frac{\partial \rho}{\partial t} + \nabla \cdot \vec{\mathbf{j}} = 0.$$

In this steady situation when things do not depend on time, we have the steady-current condition

$$\nabla \cdot \vec{\mathbf{j}} = 0.$$

We will focus on the situation $\rho = 0$ in the study of magnetostatics in this chapter.

2.5.1 Magnetic Field

The Ampère-Maxwell law

$$\nabla \times \vec{\mathbf{B}} = \mu_0 \left(\vec{\mathbf{j}} + \varepsilon_0 \frac{\partial \vec{\mathbf{E}}}{\partial t} \right)$$

is reduced to the Ampère's law in the steady case

$$\nabla \times \vec{\mathbf{B}} = \mu_0 \vec{\mathbf{j}}.$$

The equations for magnetic fields generated by a steady current are

$$\begin{cases} \nabla \times \vec{\mathbf{B}} = \mu_0 \vec{\mathbf{j}} & \text{Ampère's law} \\ \nabla \cdot \vec{\mathbf{B}} = 0 & \text{Gauss's law} \end{cases}$$

The Gauss's law implies that we can express (in the region with $H^2_{dR} = 0$)

$$\vec{\mathbf{B}} = \nabla \times \vec{\mathbf{A}}$$

in terms of a vector $\vec{\mathbf{A}}$, called the **vector potential**. This representation is not unique: the magnetic field $\vec{\mathbf{B}}$ remains the same under the gauge transformation

$$\vec{\mathbf{A}} \longmapsto \vec{\mathbf{A}}' = \vec{\mathbf{A}} + \nabla \chi$$

for a function $\chi = \chi(\vec{r})$. In terms of the vector potential, the Ampère's law becomes

$$\nabla \times (\nabla \times \vec{\mathbf{A}}) = \mu_0 \vec{\mathbf{j}}.$$

A direct calculation shows that this is equivalent to

$$-\nabla^2 \vec{\mathbf{A}} + \nabla(\nabla \cdot \vec{\mathbf{A}}) = \mu_0 \vec{\mathbf{j}}.$$

Remark 2.5.1. One quick way to see this is to use differential forms. In fact, let us identify the vector $\vec{\mathbf{A}}$ with a 1-form

$$\mathbb{A} = \mathbf{A}_x dx + \mathbf{A}_y dy + \mathbf{A}_z dz.$$

The vector $\nabla \times \vec{\mathbf{A}}$ corresponds to the 1-form $*_3 d_3 \mathbb{A}$. Thus the vector $\nabla \times (\nabla \times \vec{\mathbf{A}})$ corresponds to

$$*_3 d_3 *_3 d\mathbb{A} = d_3^* d_3 \mathbb{A}.$$

Using $d_3 d_3^* + d_3^* d_3 = -\nabla^2$, we find

$$d_3^* d_3 \mathbb{A} = -\nabla^2 \mathbb{A} - d_3 d_3^* \mathbb{A} = -\nabla^2 \mathbb{A} + d_3(\nabla \cdot \vec{\mathbf{A}})$$

which corresponds to the vector $-\nabla^2 \vec{\mathbf{A}} + \nabla(\nabla \cdot \vec{\mathbf{A}})$.

Due to the gauge degrees of freedom, we can choose some gauge fixing condition to simplify the problem. A natural choice is the

$$\textbf{Coulomb gauge}: \quad \nabla \cdot \vec{\mathbf{A}} = 0.$$

This can also be achieved for domains in question. In fact, assume $\nabla \cdot \vec{\mathbf{A}} = f \neq 0$. Consider the gauge transformation

$$\vec{\mathbf{A}}' = \vec{\mathbf{A}} + \nabla \chi.$$

Solving the Coulomb gauge condition $\nabla \cdot \vec{\mathbf{A}}' = 0$ is to solve

$$\nabla^2 \chi = -f$$

which is the Poisson's equation. With appropriate boundary condition, the solution exists and is unique.

Now let us assume the vector potential satisfies the Coulomb gauge: $\nabla \cdot \vec{\mathbf{A}} = 0$. Then the Ampère's law in such gauge becomes

$$\nabla^2 \vec{\mathbf{A}} = -\mu_0 \vec{\mathbf{j}}$$

which becomes the vector version of the Poisson's equation. Our experience in electrostatics allows us to write down the solution in the space

$$\vec{\mathbf{A}}(\vec{r}) = \frac{\mu_0}{4\pi} \int d^3 r' \, \frac{\vec{\mathbf{j}}(\vec{r}')}{|\vec{r} - \vec{r}'|}.$$

We can check that the Coulomb gauge is satisfied when the current $\vec{\mathbf{j}}$ is suitably located in some region V. Then

$$\begin{aligned} \nabla \cdot \vec{\mathbf{A}}(\vec{r}) &= \frac{\mu_0}{4\pi} \int d^3 r' \, \vec{\mathbf{j}}(\vec{r}') \cdot \nabla \left(\frac{1}{|\vec{r} - \vec{r}'|} \right) = -\frac{\mu_0}{4\pi} \int d^3 r' \, \vec{\mathbf{j}}(\vec{r}') \cdot \nabla' \left(\frac{1}{|\vec{r} - \vec{r}'|} \right) \\ &= \frac{\mu_0}{4\pi} \int d^3 r' \, \left(\nabla' \cdot \vec{\mathbf{j}}(\vec{r}') \right) \frac{1}{|\vec{r} - \vec{r}'|} = 0. \end{aligned}$$

Here we have used integration by part and the steady current condition $\nabla \cdot \vec{\mathbf{j}} = 0$.

As a result, the magnetic field can be written as

$$\vec{\mathbf{B}}(\vec{r}) = \nabla \times \vec{\mathbf{A}}(\vec{r}) = \frac{\mu_0}{4\pi} \int d^3 r' \, \vec{\mathbf{j}}(\vec{r}') \times \frac{\vec{r} - \vec{r}'}{|\vec{r} - \vec{r}'|^3}.$$

This formula is the Biot-Savart law.

There is also an integral form of Ampère's law. Consider a surface S with boundary curve $C = \partial S$. Consider the curve integral

$$\oint_C d\vec{r} \cdot \vec{\mathbf{B}} = \int_S d\vec{\sigma} \cdot (\nabla \times \vec{\mathbf{B}}) = \mu_0 \int_S d\vec{\sigma} \cdot \vec{\mathbf{j}} = \mu_0 I$$

where $I = \int_S d\vec{\sigma} \cdot \vec{\mathbf{j}}$ is the total current passing through S.

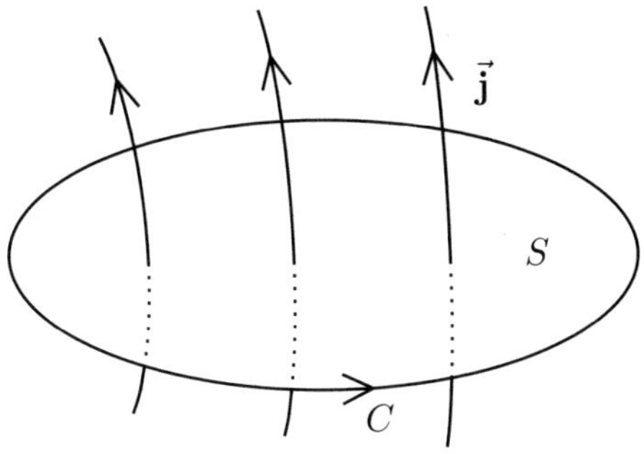

Therefore we find

$$\oint_C d\vec{r} \cdot \vec{\mathbf{B}} = \mu_0 I.$$

Example 2.5.2 (Infinite Straight Wire)**.** Consider an infinite straight wire carrying a steady current I. Assume the wire is placed along the z-axis. Symmetry leads us to consider cylindrical polar coordinate

$$(\rho, \theta, z)$$

where $\rho = \sqrt{x^2 + y^2}$. The current is

$$\vec{\mathbf{j}} = I\, \delta(x)\delta(y)\, \hat{z}.$$

Here $\hat{z}$ is the unit vector along the z-direction.

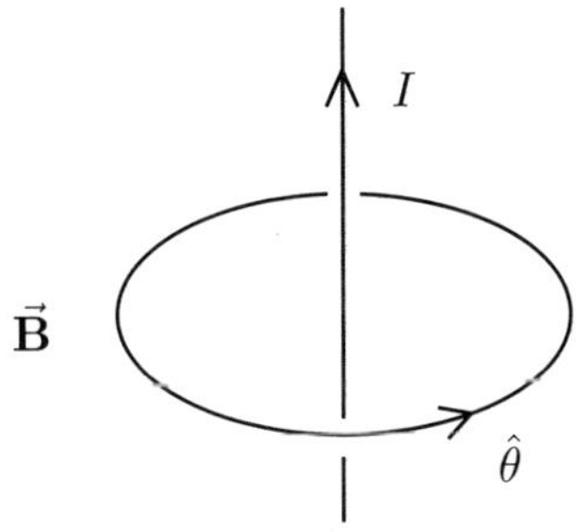

By transformation symmetry, $\vec{\mathbf{B}} = \vec{\mathbf{B}}(\rho, \theta)$ does not depend on z. Using Biot-Savart law,

$$\vec{\mathbf{B}}(\rho, \theta) = \frac{\mu_0}{4\pi} \int_{-\infty}^{\infty} dl \, \frac{I\rho}{(\rho^2 + l^2)^{3/2}} \hat{\theta} \xlongequal{l \to \rho l} \frac{\mu_0 I}{4\pi\rho} \hat{\theta} \int_{-\infty}^{\infty} \frac{dl}{(1 + l^2)^{3/2}}.$$

Here $\hat{\theta}$ is the unit vector along the θ-direction. The last integral can be performed $\int_{-\infty}^{\infty} \frac{dl}{(1+l^2)^{3/2}} = 2$. Then

$$\vec{\mathbf{B}}(\rho, \theta) = \frac{\mu_0 I}{2\pi\rho} \hat{\theta}.$$

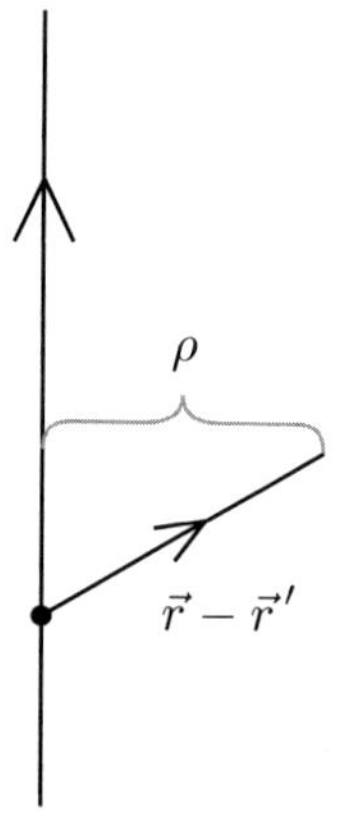

We can check this result by considering a loop integral along the circle C_ρ of radius ρ in the xy-plane centered at the origin.

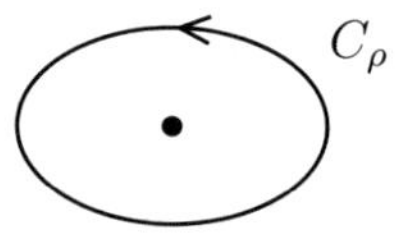

$$\oint_{C_\rho} d\vec{r} \cdot \vec{\mathbf{B}}(\rho, \theta) = \int_0^{2\pi} \rho d\theta \, \frac{\mu_0 I}{2\pi\rho} = \mu_0 I$$

as expected.

Example 2.5.3 (Infinite Plane Current). Consider the surface of xy-plane carrying a constant surface current density $\vec{K}$ per unit length. Assume $\vec{K} = k\hat{x}$ lies in the x-direction.

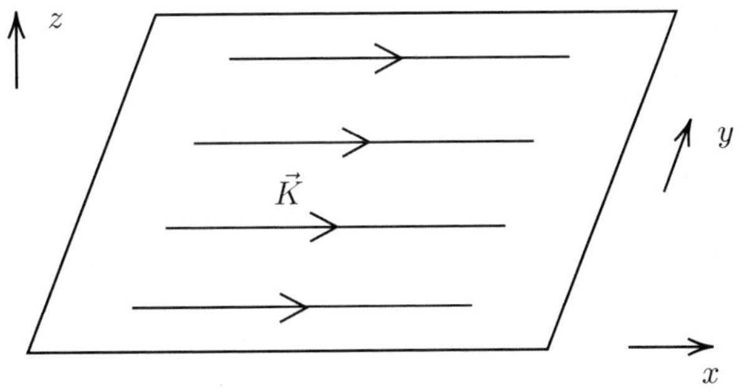

Based on our experience from the infinite line wire, and the translation symmetry, we see that $\vec{\mathbf{B}}$ is oriented along the direction of $-\hat{y}$ when $z > 0$, along the direction of $\hat{y}$ when $z < 0$, and its magnitude only depends on z. Let us write

$$\vec{\mathbf{B}} = -\mathbf{B}(z)\hat{y},$$

where $\mathbf{B}(-z) = -\mathbf{B}(z)$. Now we consider Ampère's law in the rectangle as illustrated.

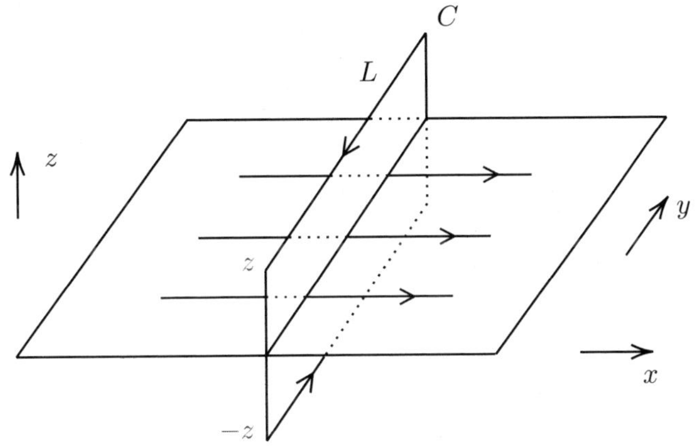

$$\oint_C d\vec{r} \cdot \vec{\mathbf{B}} = \mathbf{B}(z)L - \mathbf{B}(-z)L = 2\mathbf{B}(z)L.$$

The total current through the surface is KL. It follows that $2\mathbf{B}(z)L = \mu_0 KL$, i.e.

$$\vec{\mathbf{B}} = \begin{cases} -\dfrac{\mu_0 K}{2}\hat{y} & \text{if } z > 0 \\ \dfrac{\mu_0 K}{2}\hat{y} & \text{if } z < 0 \end{cases}$$

Note that the magnitude of the magnetic field is constant, which is the same situation as in electrostatics. The magnetic field is not continuous and exhibits a jump across a surface current. Let

$$\vec{\mathbf{B}}_{\pm} = \text{limit of the magnetic field along the } \pm\hat{n} \text{ side of the surface.}$$

Here is the example $\hat{n} = \hat{z}$. Then

$$\hat{n} \times (\vec{\mathbf{B}}_{+} - \vec{\mathbf{B}}_{-}) = \mu_0 \vec{K}.$$

2.5.2 Interface Condition

A similar argument as in our discussion of electric field interface condition in Section 2.2.3 leads to the following interface condition for magnetic field across a surface with current density $\vec{K}$ and normal vector $\hat{n}$.

$$\begin{cases} \hat{n} \cdot (\vec{\mathbf{B}}_+ - \vec{\mathbf{B}}_-) = 0 & \text{continuous along normal direction} \\ \hat{n} \times (\vec{\mathbf{B}}_+ - \vec{\mathbf{B}}_-) = \mu_0 \vec{K} & \text{jump along tangent direction} \end{cases}$$

We leave the details to the reader.

Example. Solenoid

A solenoid consists of a surface current travelling around the cylinder. We consider an infinite cylinder with surface density in the rotating direction as in the picture.

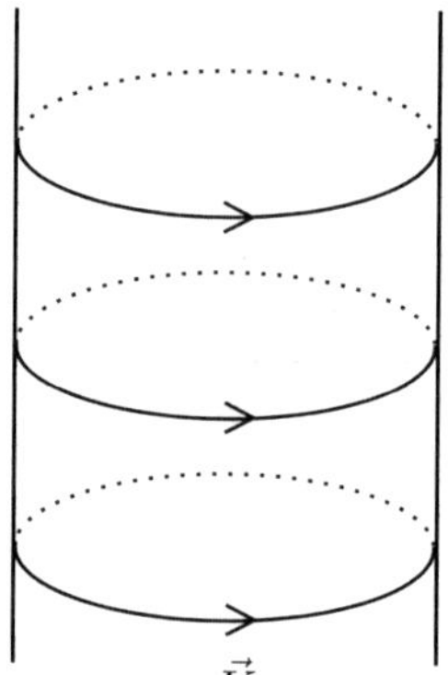

Let

$$(\rho, \theta, z)$$

be the cylinder polar coordinate. Then

$$\vec{K} = K\hat{\theta}$$

along the cylinder. By symmetry, $\vec{\mathbf{B}}$ points in the z-direction and is a function of ρ only

$$\vec{\mathbf{B}} = \mathbf{B}(\rho)\hat{z}.$$

Away from the cylinder, we have $\vec{\mathbf{j}} = 0$ and

$$\nabla \times \vec{\mathbf{B}} = 0 \quad \Longrightarrow \quad \frac{d\mathbf{B}}{d\rho} = 0 \quad \Longrightarrow \quad \mathbf{B} \text{ is const.}$$

Outside the cylinder: $\mathbf{B}(+\infty) = 0 \Longrightarrow \vec{\mathbf{B}} = 0$ outside the cylinder.

Inside the cylinder: Consider the surface

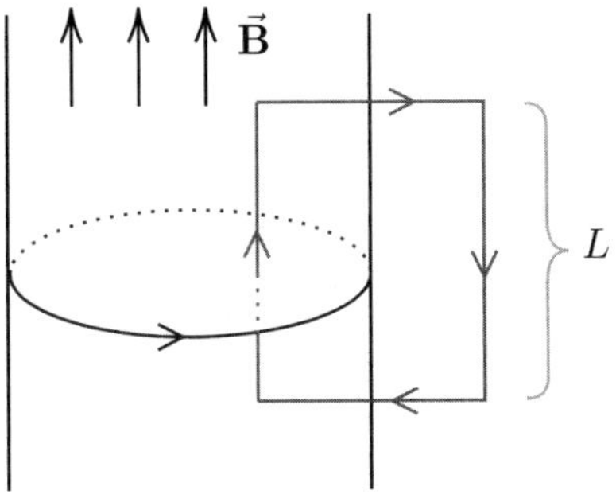

$$\oint_C d\vec{r} \cdot \vec{\mathbf{B}} = \mathbf{B}(\rho)L, \qquad I = KL \qquad \Longrightarrow \qquad \mathbf{B}(\rho) = \mu_0 K, \qquad \vec{\mathbf{B}} = \mu_0 K \hat{z}.$$

2.6 Magnetic Moment and Magnetic Dipole

2.6.1 Magnetic Moment

Assume we have a current distribution $\vec{\mathbf{j}}$ localized in a finite region V. It generates a vector potential via

$$\vec{\mathbf{A}}(\vec{r}) = \frac{\mu_0}{4\pi} \int_V d^3 r' \, \frac{\vec{\mathbf{j}}(\vec{r}\,')}{|\vec{r} - \vec{r}\,'|}.$$

In parallel with the electrostatic case, we consider an approximation to $\vec{\mathbf{A}}$ when $\vec{r}$ lies far away from the region V, called the **magnetic multipole expansion**. The idea is the same as the electrostatic case. Consider

$$\frac{1}{|\vec{r} - \vec{r}\,'|} = \frac{1}{r} \sum_{l=0}^{\infty} \left(\frac{r'}{r}\right)^l P_l(\hat{r} \cdot \hat{r}\,'), \qquad r >> 0.$$

Here $\hat{r} = \frac{\vec{r}}{r}$, and P_l's are Legendre polynomials. Plug this into the vector potential, we find

$$\begin{aligned}
\vec{\mathbf{A}}(\vec{r}) &= \frac{\mu_0}{4\pi r} \int_V d^3 r' \sum_{l=0}^{\infty} \left(\frac{r'}{r}\right)^l P_l(\hat{r} \cdot \hat{r}\,') \vec{\mathbf{j}}(\vec{r}\,') \\
&= \frac{\mu_0}{4\pi} \int_V d^3 r' \Big(\underbrace{\vec{\mathbf{j}}(\vec{r}\,')\frac{1}{r}}_{\text{monopole term}} + \underbrace{\vec{\mathbf{j}}(\vec{r}\,')\frac{\vec{r} \cdot \vec{r}\,'}{r^3}}_{\text{dipole term}} + \underbrace{\cdots}_{\text{multipole}} \Big).
\end{aligned}$$

The monopole term vanishes. For example, let us consider the x-component. Denote $\vec{r}\,' = (x', y', z')$, $\vec{\mathbf{j}} = (\mathbf{j}_x, \mathbf{j}_y, \mathbf{j}_z)$ in components. We have

$$\int_V d^3 r' \, \mathbf{j}_x(\vec{r}\,') = \int_V d^3 r' \, \nabla' \cdot (x' \vec{\mathbf{j}}) - x' \nabla' \cdot \vec{\mathbf{j}} = 0$$

since $\vec{\mathbf{j}}$ is localized inside V and $\nabla' \cdot \vec{\mathbf{j}} = 0$.

Therefore the leading contribution comes from the dipole term

$$\frac{\mu_0}{4\pi r^3}\int_V d^3r'\,(\vec{r}\cdot\vec{r}\,')\vec{\mathbf{j}}(\vec{r}\,').$$

Using $\mathbf{j}_x = \nabla'\cdot(x'\vec{\mathbf{j}}(\vec{r}\,'))$ as above, we have

$$\begin{aligned}\int_V d^3r'\,(\vec{r}\cdot\vec{r}\,')\mathbf{j}_x(\vec{r}\,') &= \int_V d^3r'\,(\vec{r}\cdot\vec{r}\,')\nabla'\cdot(x'\vec{\mathbf{j}}) = -\int_V d^3r'\,\nabla'(\vec{r}\cdot\vec{r}\,')\cdot x'\vec{\mathbf{j}}\\ &= -\int_V d^3r'\,(\vec{r}\cdot\vec{\mathbf{j}})x'.\end{aligned}$$

Similar formula holds for y, z components, so we get

$$\int_V d^3r'\,(\vec{r}\cdot\vec{r}\,')\mathbf{j}(\vec{r}\,') = -\int_V d^3r'\,(\vec{r}\cdot\vec{\mathbf{j}})\vec{r}\,'.$$

Now using the vector identity $\vec{a}\times(\vec{b}\times\vec{c}) = (\vec{a}\cdot\vec{c})\vec{b} - (\vec{a}\cdot\vec{b})\vec{c}$,

$$(\vec{r}\,'\times\vec{\mathbf{j}})\times\vec{r} = (\vec{r}\cdot\vec{r}\,')\vec{\mathbf{j}} - (\vec{r}\cdot\vec{\mathbf{j}})\vec{r}\,',$$

we find

$$\begin{aligned}\int_V d^3r'\,(\vec{r}\cdot\vec{r}\,')\vec{\mathbf{j}}(\vec{r}\,') =& \frac{1}{2}\int_V d^3r'\,\left[(\vec{r}\cdot\vec{r}\,')\vec{\mathbf{j}}(\vec{r}\,') - (\vec{r}\cdot\vec{\mathbf{j}})\vec{r}\,'\right]\\ =& \left(\frac{1}{2}\int_V d^3r'\,\vec{r}\,'\times\vec{\mathbf{j}}(\vec{r}\,')\right)\times\vec{r}.\end{aligned}$$

Define the **magnetic (dipole) moment** by

$$\vec{m} = \frac{1}{2}\int_V d^3r'\,\vec{r}\,'\times\vec{\mathbf{j}}(\vec{r}\,').$$

Then the magnetic dipole approximation is

$$\vec{\mathbf{A}}(\vec{r}) \approx \frac{\mu_0}{4\pi}\frac{\vec{m}\times\vec{r}}{r^3} = -\frac{\mu_0}{4\pi}\vec{m}\times\nabla\left(\frac{1}{r}\right).$$

The corresponding magnetic field is (when $r >> 0$)

$$\begin{aligned}\vec{\mathbf{B}} =& \nabla\times\vec{\mathbf{A}} \simeq -\frac{\mu_0}{4\pi}\nabla\times\left(\vec{m}\times\nabla\left(\frac{1}{r}\right)\right)\\ =& -\frac{\mu_0}{4\pi}\vec{m}\nabla^2\left(\frac{1}{r}\right) + \frac{\mu_0}{4\pi}(\vec{m}\cdot\nabla)\nabla\left(\frac{1}{r}\right)\\ =& -\nabla\left(\frac{\mu_0}{4\pi}\frac{\vec{m}\cdot\vec{r}}{r^3}\right).\end{aligned}$$

Comparing with the electrostatic dipole approximation

$$\phi(\vec{r}) = \frac{1}{4\pi\varepsilon_0}\frac{\vec{p}\cdot\vec{r}}{r^3}, \qquad \vec{\mathbf{E}} = -\nabla\phi,$$

we see that the magnetic dipole approximation takes the same form as that of electric dipole

$$\vec{\mathbf{B}} = \frac{\mu_0}{4\pi}\frac{3(\hat{r}\cdot\vec{m})\hat{r} - \vec{m}}{r^3}, \qquad r >> 0.$$

Example 2.6.1 (Magnetic Moment and Angular Momentum). Consider the current generated by the motion of a number of charged particles with charges q_i, masses M_i, positions $\vec{r}_i$ and velocities $\vec{v}_i = \dot{\vec{r}}_i$. The current density is

$$\vec{\mathbf{j}}(\vec{r}) = \sum_i q_i \vec{v}_i \delta(\vec{r} - \vec{r}_i).$$

The magnetic (dipole) moment is

$$\vec{m} = \frac{1}{2}\int d^3r'\, \vec{r}' \times \vec{\mathbf{j}}(\vec{r}') = \frac{1}{2}\sum_i q_i \vec{r}_i \times \vec{v}_i = \frac{1}{2}\sum_i \frac{q_i \vec{L}_i}{M_i}$$

where $\vec{L}_i = M_i \vec{r}_i \times \vec{v}_i$ is the angular momentum of the i-th particle. Assume all the particles in motion have the same charge/mass radio $q_i/M_i = q/M$. Then

$$\vec{m} = \frac{q}{2M}\vec{L}$$

is proportional to the total angular momentum $\vec{L} = \sum_i \vec{L}_i$.

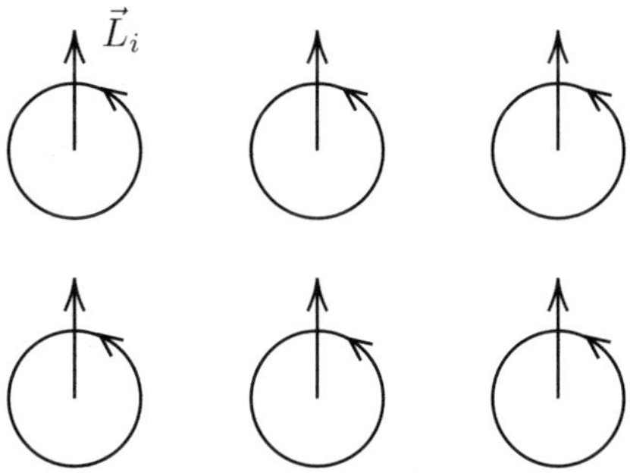

This classical result is very close to the quantum case describing magnet. The quantum particles (like electron) carry the so-called "**spin**" angular momentum $\vec{S}$, leading to an intrinsic spin magnetic moment

$$\vec{m}_g = g\frac{q}{2M}\vec{S}.$$

The number g is called the g-**factor**, and contains important information about the quantum physics.

Example 2.6.2 (Current Loop)**.** Consider a loop C carrying a steady current I.

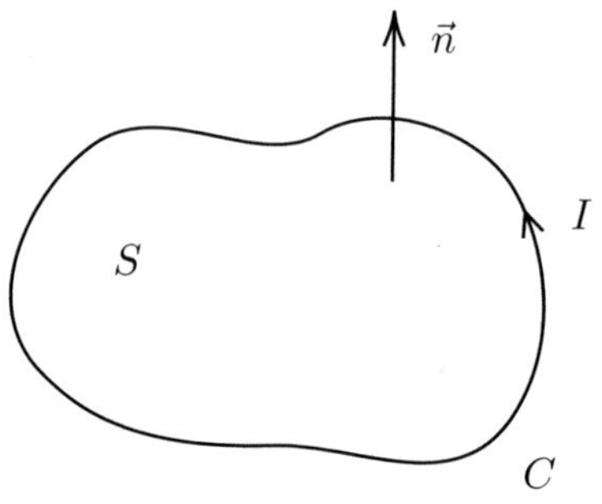

The magnetic moment is

$$\vec{m}_C = \frac{1}{2} \int d^3r \, \vec{r} \times \vec{\mathbf{j}}(\vec{r}) = \frac{I}{2} \oint_C \vec{r} \times d\vec{r}.$$

Assume the circuit C bounds a surface S. Let $\vec{\xi}$ be an arbitrary constant vector. Then

$$\begin{aligned}\vec{\xi} \cdot \vec{m}_C &= \frac{I}{2} \oint_C \vec{\xi} \cdot (\vec{r} \times d\vec{r}) = \frac{I}{2} \oint_C d\vec{r} \cdot (\vec{\xi} \times \vec{r}) = \frac{I}{2} \int_S d\vec{\sigma} \cdot \nabla \times (\vec{\xi} \times \vec{r}) \\ &= \frac{I}{2} \int_S d\vec{\sigma} \cdot \left[\vec{\xi} \nabla \cdot \vec{r} - (\vec{\xi} \cdot \nabla) \vec{r} \right] = I \int_S d\vec{\sigma} \cdot \vec{\xi}.\end{aligned}$$

Since $\vec{\xi}$ is arbitrary, we find

$$\vec{m}_C = I \int_S d\vec{\sigma}$$

which holds for any surface S with boundary C.

Now assume C is planer, say lies on a plane with normal vector $\vec{n}$ oriented as indicated. Then

$$\int_S d\vec{\sigma} = A\vec{n}$$

where A is the area circumscribed by C. In this case

$$\vec{m}_C = IA\vec{n}.$$

Now let us consider C to be a circle of radius R.

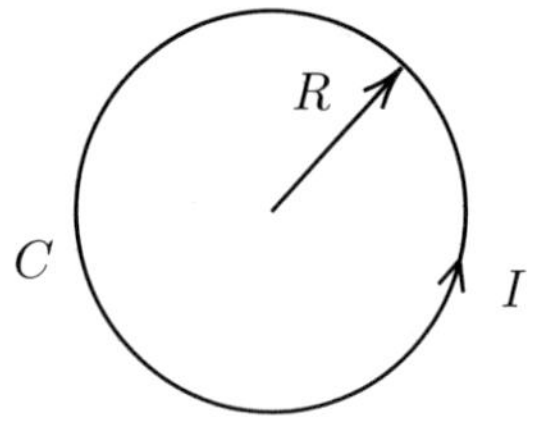

It generates a vector potential by

$$\vec{\mathbf{A}}(\vec{r}) = \frac{\mu_0 I}{4\pi} \oint_C \frac{d\vec{r}'}{|\vec{r}-\vec{r}'|} = \frac{\mu_0 I}{4\pi r} \oint_C d\vec{r}' \left[\sum_{l=0}^{\infty} \left(\frac{r'}{r}\right)^l P_l(\hat{r}\cdot\hat{r}') \right]$$

$$= \frac{\mu_0}{4\pi} \frac{\vec{m}_C \times \vec{r}}{r^3} + \frac{\mu_0 I}{4\pi r} \oint_C d\vec{r}' \left[\sum_{l\geq 2} \left(\frac{r'}{r}\right)^l P_l(\hat{r}\cdot\hat{r}') \right].$$

Here $\vec{m}_C = IA\hat{n} = \pi I R^2 \hat{n}$. Let us now consider the limit $R \to 0$, $I \to \infty$, while leaving $a = IA = \pi I R^2$ fixed. In the limit,

$$I \oint_C d\vec{r}' (\vec{r}')^l \sim R^{l-1}$$

hence all higher multipoles go to zero. In this way, we obtain a **magnetic dipole** with magnetic moment

$$\vec{m} = a\hat{n}.$$

This is similar to the limit approach to electric dipole.

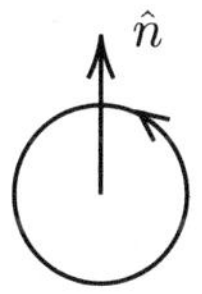

very small loop

2.6.2 Magnetic Dipole Layers

We consider a model for a surface S carrying a continuous distribution of magnetic dipoles. This is called a **magnetic dipole layer**. Let us assume that the dipoles are oriented normal to the surface, with the magnetic moment distribution given by

$$d\vec{m} = I d\vec{\sigma}.$$

Let C be the boundary of S.

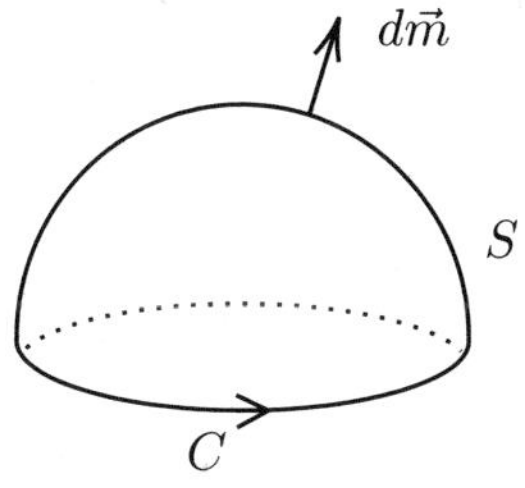

Question: What is the magnetic field produced by such a magnetic dipole layer?

Let us first consider an intuitive approach. We can think of a magnetic dipole as a small current loop

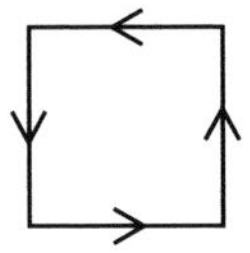

and decompose the surface S into small current loops

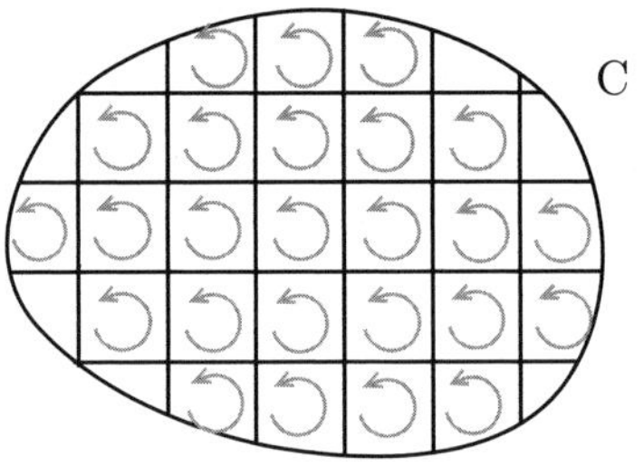

The currents inside S will cancel each other, leading to the effect of a total current I circling along C. As a result, we should expect that the magnetic field should be the same as that produced by current I circling along the loop C.

Let us now confirm this intuitive result. By the magnetic dipole formula, we have

$$\vec{\mathbf{A}}(\vec{r}) = \frac{\mu_0}{4\pi}\int_S d\vec{m} \times \frac{\vec{r}-\vec{r}'}{|\vec{r}-\vec{r}'|^3} = \frac{\mu_0}{4\pi}\int_S I d\vec{\sigma} \times \nabla'\left(\frac{1}{|\vec{r}-\vec{r}'|}\right).$$

Here the position on the surface S is parametrized by $\vec{r}'$. Let us choose again an arbitrary constant vector $\vec{\xi}$, then

$$\begin{aligned}\vec{\xi}\cdot\vec{\mathbf{A}}(\vec{r}) &= \frac{\mu_0 I}{4\pi}\int_S \vec{\xi}\cdot\left(d\vec{\sigma}\times\nabla'\left(\frac{1}{|\vec{r}-\vec{r}'|}\right)\right) = \frac{\mu_0 I}{4\pi}\int_S d\vec{\sigma}\cdot\nabla'\times\frac{\vec{\xi}}{|\vec{r}-\vec{r}'|}\\ &= \frac{\mu_0 I}{4\pi}\oint_C d\vec{r}'\cdot\frac{\vec{\xi}}{|\vec{r}-\vec{r}'|} = \vec{\xi}\cdot\left(\frac{\mu_0 I}{4\pi}\oint_C\frac{d\vec{r}'}{|\vec{r}-\vec{r}'|}\right).\end{aligned}$$

This holds for any vector $\vec{\xi}$, hence

$$\vec{\mathbf{A}}(\vec{r}) = \frac{\mu_0 I}{4\pi}\oint_C \frac{d\vec{r}'}{|\vec{r}-\vec{r}'|}$$

which is indeed the vector potential of a loop C carrying a current I.

2.7 Linking and Magnetic Helicity

We discuss some basic idea about topological aspects of magnetic field. The static Maxwell's equations

$$\begin{cases} \nabla \cdot \vec{\mathbf{B}} = 0 & \text{Gauss's law} \\ \nabla \times \vec{\mathbf{B}} = \mu_0 \vec{\mathbf{j}} & \text{Ampère's law} \end{cases}$$

have the integral form

$$\begin{cases} \displaystyle\int_{\partial V} d\vec{\sigma} \cdot \vec{\mathbf{B}} = 0 \\ \displaystyle\oint_{\partial S} d\vec{r} \cdot \vec{\mathbf{B}} = \mu_0 \int_S d\vec{\sigma} \cdot \vec{\mathbf{j}} = \mu_0 I \end{cases}$$

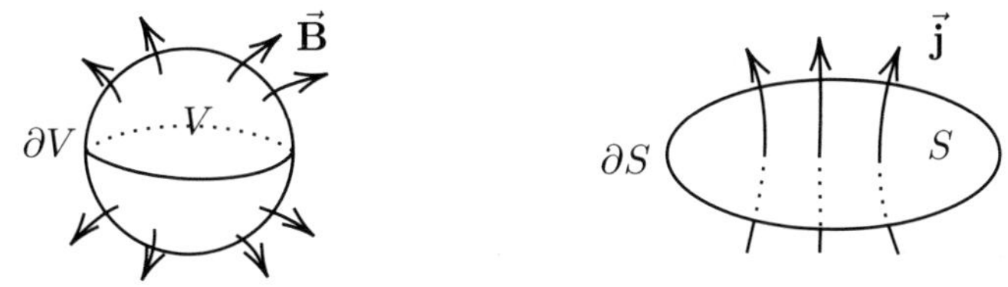

Here V is a region and S is a surface in $\mathbb{R}^3$. ∂V and ∂S are their boundaries.

The Biot-Savart law gives a formula of $\vec{\mathbf{B}}$ from the current $\vec{\mathbf{j}}$. One interesting consequence is the Gauss's linking number formula that we have seen before. Let us briefly recall here. Consider a circle C_2 carrying a steady current I. It generates a magnetic field as illustrated

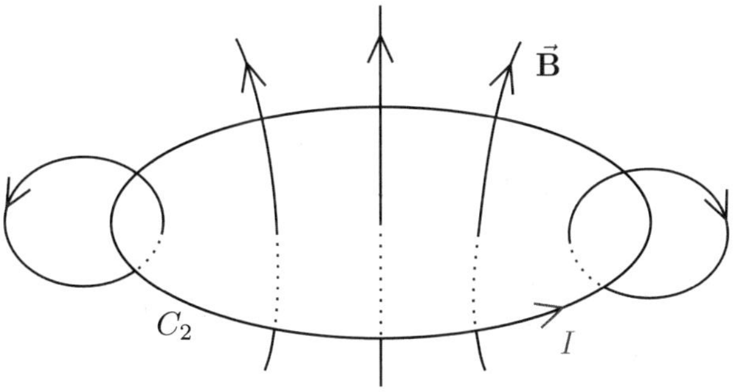

Consider another circle C_1 that links with C_2 as before. The total current through S is

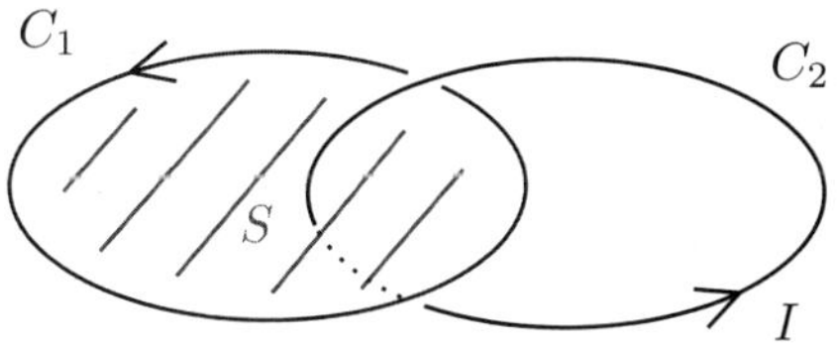

$n(C_1, C_2)I$ where $n(C_1, C_2)$ is the linking number of C_1 and C_2. Then

$$\oint_{C_1} d\vec{r}_1 \cdot \vec{\mathbf{B}}(\vec{r}_1) = \mu_0 n(C_1, C_2) I$$

together with the Biot-Savart law

$$\vec{\mathbf{B}}(\vec{r}_1) = \frac{\mu_0 I}{4\pi} \oint_{C_2} d\vec{r}_2 \times \frac{\vec{r}_1 - \vec{r}_2}{|\vec{r}_1 - \vec{r}_2|^3}$$

leads to the Gauss's integral formula

$$n(C_1, C_2) = \frac{1}{4\pi} \oint_{C_1} d\vec{r}_1 \cdot \oint_{C_2} d\vec{r}_2 \times \frac{\vec{r}_1 - \vec{r}_2}{|\vec{r}_1 - \vec{r}_2|^3}.$$

There is a way to capture the topological complexity of magnetic field based on similar idea, called the **magnetic helicity**. It is defined by

$$H = \int_V d^3r\, \vec{\mathbf{A}} \cdot \vec{\mathbf{B}}.$$

In terms of differential forms, this is (Exercise)

$$H = \int_V \mathbb{A} \wedge d\mathbb{A}$$

where $\mathbb{A} = \mathbf{A}_x dx + \mathbf{A}_y dy + \mathbf{A}_z dz$. For those who are familiar with geometry, you realize that this is the same form of **abelian Chern-Simons action**.

Let us first consider the gauge invariance of the magnetic helicity. Under the gauge transformation

$$\mathbb{A} \longmapsto \mathbb{A} + d\chi,$$

the magnetic helicity changes by

$$\int_V d\chi \wedge d\mathbb{A} = \int_{\partial V} \chi d\mathbb{A} = \int_{\partial V} \chi *_3 \mathbb{B} = \int_{\partial V} \chi\, (\hat{n} \cdot \vec{\mathbf{B}})$$

where $\hat{n}$ is the normal vector on the surface ∂V. We see that in the case when ∂V is a magnetic surface ($\hat{n} \cdot \vec{\mathbf{B}} = 0$), the magnetic helicity will be invariant under gauge transformations.

Let us use Biot-Savart law to analyze the magnetic helicity. We assume the current $\vec{\mathbf{j}}$ lies inside V and ∂V is a magnetic surface. Then the vector potential is given by

$$\vec{\mathbf{A}}(\vec{r}) = \frac{\mu_0}{4\pi} \int_V d^3r' \frac{\vec{\mathbf{j}}(\vec{r}')}{|\vec{r} - \vec{r}'|}.$$

Using Ampère's law $\nabla \times \vec{\mathbf{B}} = \mu_0 \vec{\mathbf{j}}$, this is

$$\vec{\mathbf{A}}(\vec{r}) = \frac{1}{4\pi} \int_V d^3r' \frac{\nabla' \times \vec{\mathbf{B}}(\vec{r}')}{|\vec{r} - \vec{r}'|} = -\frac{1}{4\pi} \int_V d^3r' \frac{(\vec{r} - \vec{r}') \times \vec{\mathbf{B}}(\vec{r}')}{|\vec{r} - \vec{r}'|^3}.$$

Then the magnetic helicity becomes

$$H = \int_V d^3r\, \vec{\mathbf{A}} \cdot \vec{\mathbf{B}} = \frac{1}{4\pi} \int_V d^3r \int_V d^3r' \, \frac{\vec{\mathbf{B}}(\vec{r}) \cdot \left(\vec{\mathbf{B}}(\vec{r}\,') \times (\vec{r} - \vec{r}\,') \right)}{|\vec{r} - \vec{r}\,'|^3}.$$

Comparing with the Gauss's linking formula, we see that the magnetic helicity can be intuitively interpreted as the averaged linking over all pairs of magnetic field lines.

2.8 Dirac Monopole

The Gauss's law of magnetic field can be written as

$$d_3 *_3 \mathbb{B} = 0$$

where $\mathbb{B} = \mathbf{B}_x dx + \mathbf{B}_y dy + \mathbf{B}_z dz$. Here recall d_3 denotes the de Rham differential on $\mathbb{R}^3$. In other words, $*_3\mathbb{B}$ is a closed 2-form, a fact which inherits many topological natures. When the region has trivial topology with $H^2_{dR} = 0$, we can always find a vector potential $\vec{\mathbf{A}}$ such that

$$*_3\mathbb{B} = d_3\mathbb{A}$$

where $\mathbb{A} = \mathbf{A}_x dx + \mathbf{A}_y dy + \mathbf{A}_z dz$. However, when we work with a region V that has nontrivial topology, we may not be able to find such an $\mathbb{A}$ globally, although locally it always exists (Poincaré's Lemma). We will systematically study this later in Chapter 4 in terms of fiber bundle. Here we illustrate some basic feature via the example of Dirac monopole.

The Dirac monopole has magnetic field

$$\vec{\mathbf{B}} = \frac{g}{4\pi} \frac{\vec{r}}{r^3}$$

where g is the magnetic charge. It satisfies

$$\nabla \cdot \vec{\mathbf{B}} = g\delta(r)$$

hence describing a magnetic monopole of charge g at the origin. Although we have not observed magnetic monopole in the lab yet, this is still an interesting theoretical model. We can compare the above formula with the point electric charge

$$\vec{\mathbf{E}} = \frac{q}{4\pi\varepsilon_0} \frac{\vec{r}}{r^3}.$$

Let us consider the region of the complement of the origin in $\mathbb{R}^3$:

$$V = \mathbb{R}^3 - \{0\}.$$

We have

$$*_3\mathbb{B} = \frac{g}{4\pi r^3} (x dy \wedge dz + y dz \wedge dx + z dx \wedge dy)$$

which defines a closed 2-form on V. However, there does not exist globally a 1-form $\mathbb{A}$ on V such that

$$*_3\mathbb{B} = d_3\mathbb{A}.$$

Indeed, assume this equation holds. Consider the surface S of the unit sphere. On one hand, we have

$$\int_S *_3\mathbb{B} = \int_S d\vec{\sigma}\cdot\vec{\mathbf{B}} = \int d^3r\nabla\cdot\vec{\mathbf{B}} = g.$$

On the other hand,

$$\int_S d\mathbb{A} = \int_{\partial S}\mathbb{A} = 0.$$

This is a contradiction.

In fact, V has the homotopy type of S^2, and

$$H^2_{\mathrm{dR}}(V) \simeq H^2_{\mathrm{dR}}(S^2) \simeq \mathbb{R}.$$

The 2-form $*_3\mathbb{B}$ defines a nontrivial element in $H^2_{\mathrm{dR}}(V)$. Instead, it is possible to write $*_3\mathbb{B}$ as $d\mathbb{A}$ locally. For example, let us consider

$$\mathcal{U}_+ = \mathbb{R}^3\setminus\{(0,0,z)\,|\,z\le 0\}$$
$$\mathcal{U}_- = \mathbb{R}^3\setminus\{(0,0,z)\,|\,z\ge 0\}$$

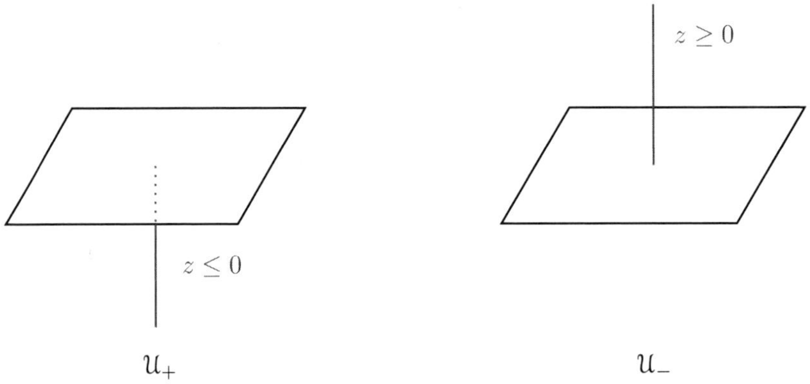

It is clear that

$$\begin{cases}\mathcal{U}_+\cup\mathcal{U}_- = V = \mathbb{R}^3 - \{0\}\\ \mathcal{U}_+\cap\mathcal{U}_- = \mathbb{R}^3 - \{z\text{-axis}\}\end{cases}$$

Consider the following 1-form on each patch

$$\mathbb{A}_\pm = \frac{g}{4\pi r(z\pm r)}(-ydx + xdy) \qquad \text{on } \mathcal{U}_\pm.$$

In spherical coordinate (r,θ,φ),

$$\begin{cases}x = r\sin\theta\cos\varphi\\ y = r\sin\theta\sin\varphi\\ z = r\cos\varphi\end{cases}$$

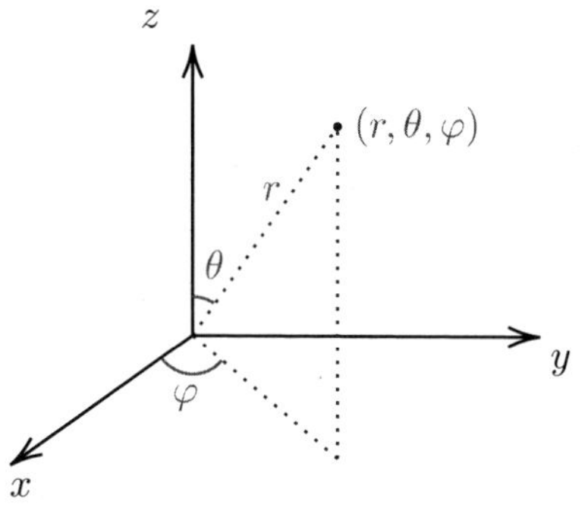

We can write

$$\mathbb{A}_{\pm} = \pm \frac{g}{4\pi}(1 \mp \cos\theta)d\varphi.$$

Note that $*_3\mathbb{B}$ in spherical coordinate is

$$*_3\mathbb{B} = \frac{g}{4\pi}\sin\theta d\theta d\varphi$$

where $\sin\theta d\theta d\varphi$ is the standard area form on S^2. It can now be easily checked that

$$*_3\mathbb{B} = d\mathbb{A}_{\pm} \qquad \text{valid on } \mathfrak{U}_{\pm}.$$

On the intersection $\mathfrak{U}_+ \cap \mathfrak{U}_-$, where both $\mathbb{A}_+$ and $\mathbb{A}_-$ are defined, we have

$$d(\mathbb{A}_+ - \mathbb{A}_-) = 0$$

so $\mathbb{A}_+ - \mathbb{A}_-$ is a closed 1-form. Explicitly, we have

$$\mathbb{A}_+ - \mathbb{A}_- = \frac{g}{2\pi}d\varphi$$

which defines a nontrivial element of

$$H^1_{\mathrm{dR}}(\mathfrak{U}_+ \cap \mathfrak{U}_-) = H^1_{\mathrm{dR}}(\mathbb{R}^3 - \{z\text{-axis}\}) \simeq H^1_{\mathrm{dR}}(S^1) = \mathbb{R}.$$

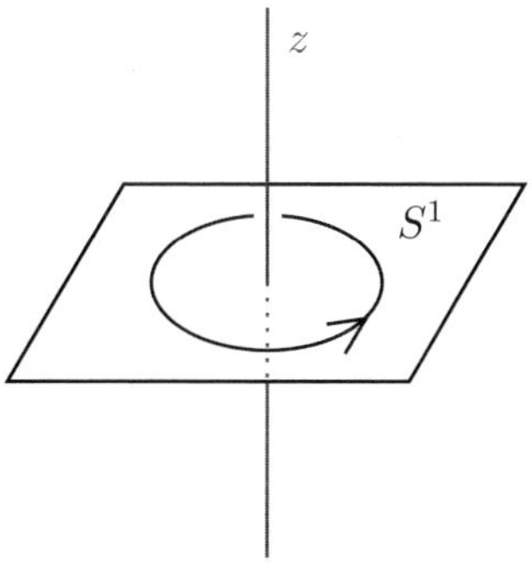

$$\mathfrak{U}_+ \cap \mathfrak{U}_- \simeq \mathbb{R}^2 - \{0\} \simeq S^1$$

Indeed, consider the unit circle S^1 on the xy-plane. Then

$$\int_{S^1} \mathbb{A}_+ - \mathbb{A}_- = \frac{g}{2\pi}\int_{S^1} d\varphi = g \neq 0.$$

An important consequence of this calculation is the Dirac quantization condition. The combination $e\mathbb{A}$ plays the role of a connection 1-form for a $U(1)$-bundle (see Section 4.6). Then the 2-form

$$\frac{1}{2\pi}ed\mathbb{A} = \frac{1}{2\pi}e{*_3}\mathbb{B}$$

is the corresponding first Chern class hence an integral form. In particular, the integration

$$\frac{1}{2\pi}\int_{S^2} e{*_3}\mathbb{B} = \frac{ge}{2\pi} \in \mathbb{Z}$$

must be an integer (Theorem 4.3.6). This gives the **Dirac quantization condition**

$$ge \in 2\pi\mathbb{Z}.$$

Chapter 3

Electrodynamics

In this chapter, we study dynamical aspects of electromagnetism where the electric and magnetic fields will evolve with time in general. We will see how electromagnetic waves arise from solving Maxwell's equations, and how they are produced by accelerating charges.

3.1 Force and Energy

3.1.1 Lorentz Force

A particle of charge q moving with velocity $\vec{v}$ in the background of electric field $\vec{\mathbf{E}}$ and magnetic field $\vec{\mathbf{B}}$ experiences force via the **Lorentz force law**

$$\vec{F} = q(\vec{\mathbf{E}} + \vec{v} \times \vec{\mathbf{B}}).$$

Example 3.1.1 (Force between steady charges). Consider two charge distributions ρ_1 and ρ_2 in the space.

Then electric field produced by ρ_2 is

$$\vec{\mathbf{E}}_2(\vec{r}) = \frac{1}{4\pi\varepsilon_0} \int d^3r' \, \rho_2(\vec{r}\,') \frac{\vec{r} - \vec{r}\,'}{|\vec{r} - \vec{r}\,'|^3}.$$

The electric force that ρ_2 exerts on ρ_1 is then given by

$$\vec{F}_1^e = \frac{1}{4\pi\varepsilon_0} \int d^3r \int d^3r' \, \rho_1(\vec{r})\rho_2(\vec{r}\,') \frac{\vec{r} - \vec{r}\,'}{|\vec{r} - \vec{r}\,'|^3}.$$

By symmetry, the force that ρ_1 exerts on ρ_2 is

$$\vec{F}_2^e = -\vec{F}_1^e.$$

In particular, $\vec{F}_1^e = \vec{F}_2^e = 0$ when $\rho_1 = \rho_2$. Thus the net force of a charge distribution on itself is zero.

Example 3.1.2 (Force between steady currents). Consider two steady current $\vec{\mathbf{j}}_1$ and $\vec{\mathbf{j}}_2$ in the space.

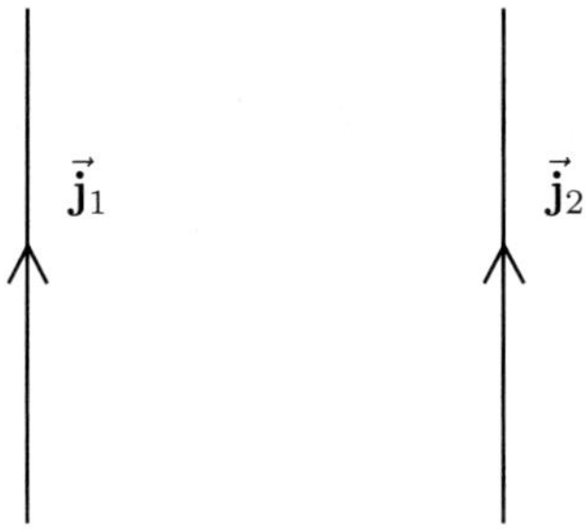

The magnetic field produced by $\vec{\mathbf{j}}_2$ is

$$\vec{\mathbf{B}}_2(\vec{r}) = \frac{\mu_0}{4\pi}\int d^3r'\,\vec{\mathbf{j}}_2(\vec{r}\,') \times \frac{\vec{r}-\vec{r}\,'}{|\vec{r}-\vec{r}\,'|^3}.$$

The magnetic force that $\vec{\mathbf{j}}_2$ exerts on $\vec{\mathbf{j}}_1$ is given by

$$\begin{aligned}\vec{F}_1^m &= \int d^3r\,\vec{\mathbf{j}}_1(\vec{r}) \times \vec{\mathbf{B}}_2(\vec{r}) = \frac{\mu_0}{4\pi}\int d^3r \int d^3r'\,\vec{\mathbf{j}}_1(\vec{r}) \times \left(\vec{\mathbf{j}}_2(\vec{r}\,') \times \frac{\vec{r}-\vec{r}\,'}{|\vec{r}-\vec{r}\,'|^3}\right) \\ &= \frac{\mu_0}{4\pi}\int d^3r \int d^3r'\,\frac{\left(\vec{\mathbf{j}}_1(\vec{r})\cdot(\vec{r}-\vec{r}\,')\right)\vec{\mathbf{j}}_2(\vec{r}\,') - \left(\vec{\mathbf{j}}_1(\vec{r})\cdot\vec{\mathbf{j}}_2(\vec{r}\,')\right)(\vec{r}-\vec{r}\,')}{|\vec{r}-\vec{r}\,'|^3}.\end{aligned}$$

In the first term, using $\nabla\cdot\vec{\mathbf{j}}_1 = 0$,

$$\int d^3r\,\vec{\mathbf{j}}_1(\vec{r})\cdot\frac{\vec{r}-\vec{r}\,'}{|\vec{r}-\vec{r}\,'|^3} = -\int d^3r\,\nabla\cdot\left(\frac{\vec{\mathbf{j}}_1(\vec{r})}{|\vec{r}-\vec{r}\,'|}\right).$$

If $\vec{\mathbf{j}}_1(\vec{r}) \to 0$ faster than $\frac{1}{r}$ as $r \to \infty$, then the divergence theorem implies that this integral vanishes. We assume this is the case. Then we find

$$\vec{F}_1^m = -\frac{\mu_0}{4\pi}\int d^3r \int d^3r'\,\left(\vec{\mathbf{j}}_1(\vec{r})\cdot\vec{\mathbf{j}}_2(\vec{r}\,')\right)\frac{\vec{r}-\vec{r}\,'}{|\vec{r}-\vec{r}\,'|^3}.$$

By symmetry, the force that $\vec{\mathbf{j}}_1$ exerts on $\vec{\mathbf{j}}_2$ is

$$\vec{F}_2^m = -\vec{F}_1^m.$$

Again, $\vec{F}_1^m = \vec{F}_2^m = 0$ when $\vec{\mathbf{j}}_1 = \vec{\mathbf{j}}_2$. Thus the net force of a current distribution on itself is zero.

Comparing with the electric force case, we find that $\vec{F}_1^m$ takes the same form as that of $\vec{F}_2^m$, except with an extra minus sign. This sign difference says that parallel currents attract and anti-parallel currents repel.

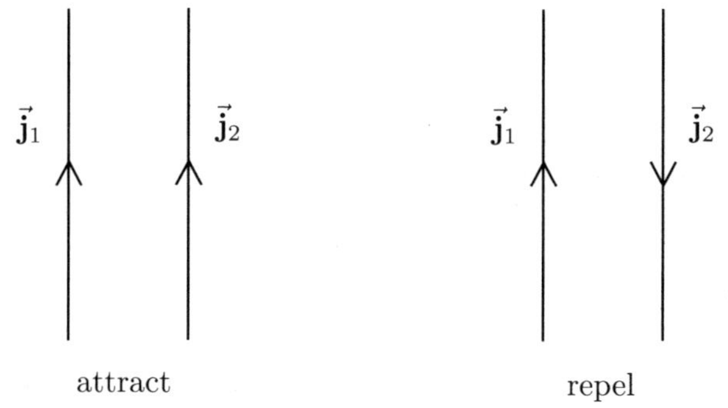

Example 3.1.3 (Helical motion in magnetic field)**.** Consider a uniform constant magnetic field

$$\vec{\mathbf{B}} = (0, 0, \mathbf{B})$$

pointing in the z-direction with magnitude $\mathbf{B}$. We consider the motion of a particle of charge q, mass m. The equation of motion is

$$m\ddot{\vec{r}} = q\dot{\vec{r}} \times \vec{\mathbf{B}}.$$

In components, this reads

$$\begin{cases} \ddot{x} = \omega_c \dot{y} \\ \ddot{y} = -\omega_c \dot{x} \\ \ddot{z} = 0 \end{cases} \qquad \text{where } \omega_c = \frac{q\mathbf{B}}{m}$$

This can be solved by

$$\begin{cases} x(t) = x_0 + R\cos(\omega_c t - \theta_0) \\ y(t) = y_0 - R\sin(\omega_c t - \theta_0) \\ z(t) = z_0 + v_z t \end{cases}$$

Here $x_0, y_0, z_0, R, \theta_0, v_z$ are constants. The particle executes circular motion at the cyclotron frequency ω_c in the xy-direction, and moves with constant speed in the z-direction.

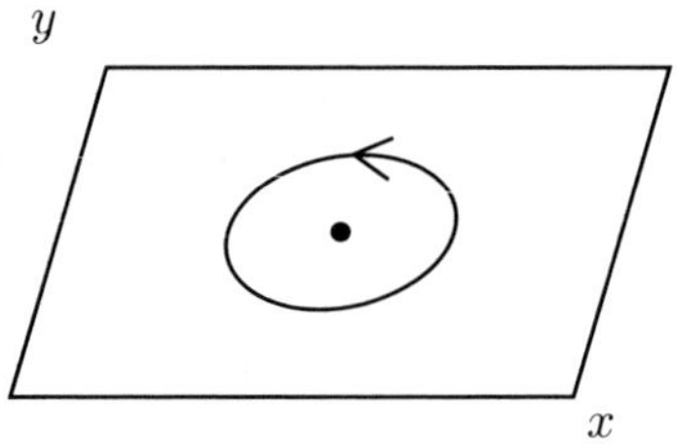

$v_z = 0$

3.1.2 Electromagnetic Energy

We explore the energy stored in electromagnetic fields. Suppose we have some charge and current configuration which produces electric field $\vec{\mathbf{E}}$ and magnetic field $\vec{\mathbf{B}}$. Here we consider the dynamical case and things will vary with time t. Then we need the full Maxwell's equations.

Consider the infinitesimal work dW done by the electromagnetic forces on these charges in the infitinesimal time dt. According to Lorentz force law, the work on a charge q is

$$\vec{F} \cdot d\vec{r} = q\left(\vec{\mathbf{E}} + \dot{\vec{r}} \times \vec{\mathbf{B}}\right) \cdot \dot{\vec{r}} dt = q\left(\vec{\mathbf{E}} \cdot \vec{v}\right) dt.$$

This can be generalized to the case of charge and current densities, by replacing

$$q \longrightarrow \rho d^3 r, \qquad \rho\vec{v} \longrightarrow \vec{\mathbf{j}}.$$

Therefore the total work done in the region V is

$$\frac{dW}{dt} = \int_V d^3 r\, \vec{\mathbf{E}} \cdot \vec{\mathbf{j}}.$$

Using Ampère-Maxwell law

$$\nabla \times \vec{\mathbf{B}} = \mu_0 \vec{\mathbf{j}} + \frac{1}{c^2}\frac{\partial \vec{\mathbf{E}}}{\partial t} = \mu_0\left(\vec{\mathbf{j}} + \varepsilon_0 \frac{\partial \vec{\mathbf{E}}}{\partial t}\right)$$

and Faraday's law

$$\nabla \times \vec{\mathbf{E}} = -\frac{\partial \vec{\mathbf{B}}}{\partial t},$$

we find

$$\begin{aligned}
\vec{\mathbf{E}} \cdot \vec{\mathbf{j}} &= \frac{1}{\mu_0}\vec{\mathbf{E}} \cdot (\nabla \times \vec{\mathbf{B}}) - \varepsilon_0 \vec{\mathbf{E}} \cdot \frac{d\vec{\mathbf{E}}}{dt} \\
&= \frac{1}{\mu_0}\left[\nabla \cdot (\vec{\mathbf{B}} \times \vec{\mathbf{E}}) + \vec{\mathbf{B}} \cdot (\nabla \times \vec{\mathbf{E}})\right] - \varepsilon_0 \vec{\mathbf{E}} \cdot \frac{d\vec{\mathbf{E}}}{dt} \\
&= -\varepsilon_0 \vec{\mathbf{E}} \cdot \frac{d\vec{\mathbf{E}}}{dt} - \frac{1}{\mu_0}\vec{\mathbf{B}} \cdot \frac{\partial \vec{\mathbf{B}}}{\partial t} - \frac{1}{\mu_0}\nabla \cdot (\vec{\mathbf{E}} \times \vec{\mathbf{B}}).
\end{aligned}$$

It follows that

$$\frac{dW}{dt} = -\frac{d}{dt}\int_V \frac{1}{2}\left(\varepsilon_0 \vec{\mathbf{E}}^2 + \frac{1}{\mu_0}\vec{\mathbf{B}}^2\right) - \frac{1}{\mu_0}\int_{\partial V} d\vec{\sigma}\cdot(\vec{\mathbf{E}}\times\vec{\mathbf{B}}).$$

This formula has the following interpretation: the quantity

$$U = \frac{1}{2}\int_V \varepsilon_0 \vec{\mathbf{E}}^2 + \frac{1}{\mu_0}\vec{\mathbf{B}}^2$$

is the **total energy** of electromagnetic field stored in the region V. Define the

$$\textbf{Poynting vector}: \quad \vec{S} = \frac{1}{\mu_0}\vec{\mathbf{E}}\times\vec{\mathbf{B}}$$

which describes the energy flow. Then

$$\frac{1}{\mu_0}\int_{\partial V} d\vec{\sigma}\cdot(\vec{\mathbf{E}}\times\vec{\mathbf{B}}) = \int_{\partial V} d\vec{\sigma}\cdot\vec{S}$$

describes the energy that flows out through ∂V.

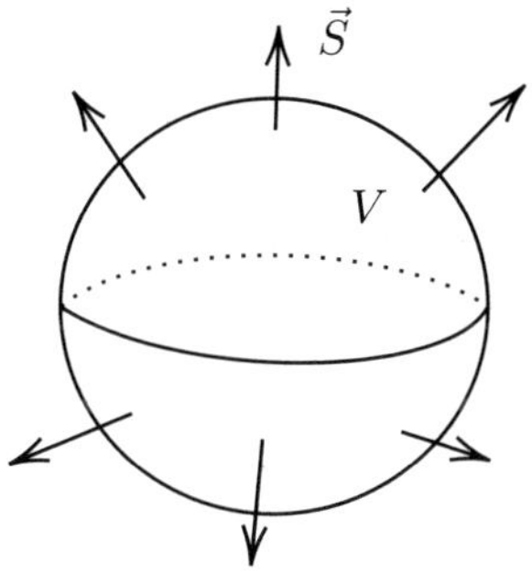

Then the formula

$$\frac{dW}{dt} = -\frac{dU}{dt} - \int_{\partial V} d\vec{\sigma}\cdot\vec{S}$$

gives the expected interpretation: the work done by the electromagnetic force is equal to the decrease in the total energy in the region less the energy that flowed out of the region. Let

$$u = \frac{1}{2}\left(\varepsilon_0 \vec{\mathbf{E}}^2 + \frac{1}{\mu_0}\vec{\mathbf{B}}^2\right)$$

denote the energy density. We have

$$\vec{\mathbf{E}}\cdot\vec{\mathbf{j}} = -\frac{\partial u}{\partial t} - \nabla\cdot\vec{S}.$$

In the region of empty space where no work is done, the above equation becomes

$$\frac{\partial u}{\partial t} + \nabla\cdot\vec{S} = 0.$$

Comparing with the charge conservation $\frac{\partial\rho}{\partial t} + \nabla\cdot\vec{\mathbf{j}} = 0$, the above equation can be viewed as the local form of the conservation of electromagnetic energy.

Example 3.1.4. Consider a uniformly charged spherical shell of total charge Q and radius R.

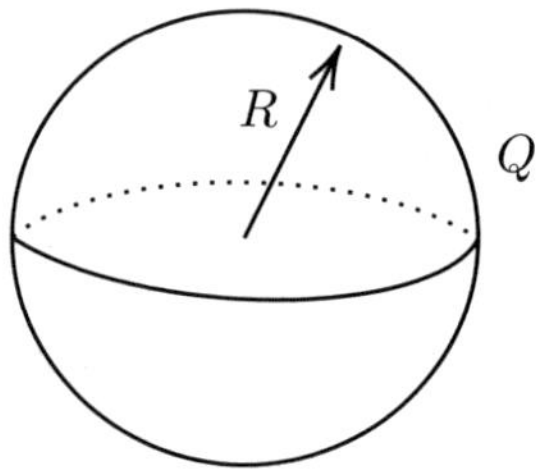

The electric field is

$$\vec{\mathbf{E}} = \begin{cases} \dfrac{Q}{4\pi\varepsilon_0}\dfrac{\vec{r}}{r^3} & r > R \\ 0 & r < R \end{cases}$$

Therefore the total energy is

$$U=\frac{\varepsilon_0}{2}\int d^3r\, \vec{\mathbf{E}}\cdot\vec{\mathbf{E}}=\frac{\varepsilon_0}{2}\int_{|\vec{r}|\geq R} d^3r \left(\frac{Q}{4\pi\varepsilon_0}\right)^2\frac{1}{r^4}=\frac{Q^2}{32\pi^2\varepsilon_0}\int_R^\infty dr(4\pi r^2)\frac{1}{r^4}=\frac{Q^2}{8\pi\varepsilon_0 R}.$$

Example 3.1.5 (Electromagnetic wave)**.** Consider the following solution of Maxwell's equations in the vacuum:

$$\begin{cases} \vec{\mathbf{E}} = \cos(\vec{k}\cdot\vec{r}-\omega t)\vec{\mathbf{E}}_0 \\ \vec{\mathbf{B}} = \dfrac{1}{c}\cos(\vec{k}\cdot\vec{r}-\omega t)\hat{k}\times\vec{\mathbf{E}}_0 \end{cases} \qquad \text{where } \hat{k} = \frac{\vec{k}}{|\vec{k}|}.$$

Here $\vec{\mathbf{E}}_0, \vec{k}, \omega$ are constants and satisfy

$$c^2\vec{k}\cdot\vec{k} = \omega^2, \qquad \vec{k}\cdot\vec{\mathbf{E}}_0 = 0.$$

This describes an electromagnetic wave that we will discuss in detail in Section 3.3. This wave travels in the direction of $\vec{k}$, while the electric field and magnetic field are oscillating in orthogonal directions.

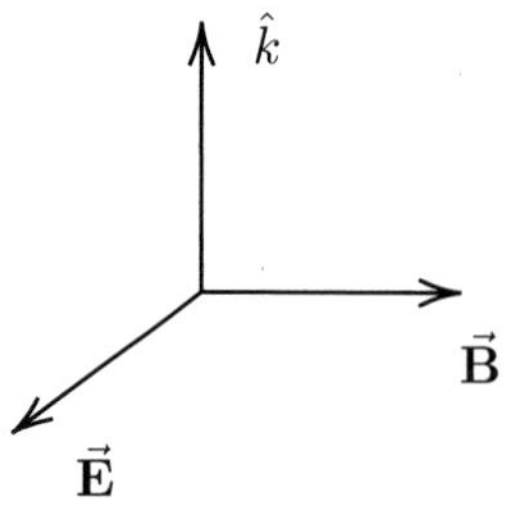

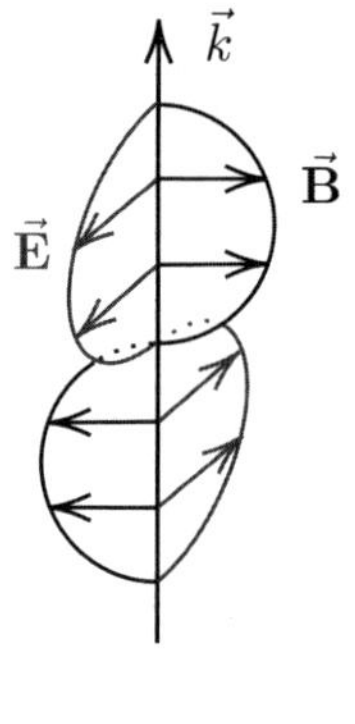

The Poynting vector is

$$\vec{S} = \frac{1}{\mu_0}\vec{\mathbf{E}} \times \vec{\mathbf{B}} = \frac{\mathbf{E}_0^2}{c\mu_0}\cos^2(\vec{k}\cdot\vec{r} - \omega t)\hat{k} \qquad \text{where } \mathbf{E}_0 = |\vec{\mathbf{E}}_0|.$$

We see that the energy is indeed transporting in the direction of the wave propagation $\hat{k}$. The energy density is

$$u = \frac{1}{2}\left(\varepsilon_0\vec{\mathbf{E}}^2 + \frac{1}{\mu_0}\vec{\mathbf{B}}^2\right) = \varepsilon_0\mathbf{E}_0^2\cos^2(\vec{k}\cdot\vec{r} - \omega t).$$

We see that

$$\vec{S} = cu\,\hat{k}$$

which simply says that the wave is transporting energy in the speed of light.

3.2 Electromagnetic Induction

We explore a bit about Faraday's law

$$\nabla \times \vec{\mathbf{E}} + \frac{\partial\vec{\mathbf{B}}}{\partial t} = 0$$

which connects the production of electric field with the change of magnetic field. This phenomenon is also called **Faraday's Law of Induction**.

3.2.1 Electromotive Force and Flux Rule

Let C be a conducting circuit. The **electromotive force** (“**emf**” for short) is the accumulated tangential force per unit charge throughout the circuit. Equivalently, emf is the work done on a unit charge moving around the whole circuit.

In a circuit context, the emf will drive the charges in the wire and form a current. In many circumstances, the resulting current I is proportional to the applied emf $\mathcal{E}$, which is called **Ohm's Law**:

$$\mathcal{E} = IR.$$

The constant of proportionality R is called the **resistance**.

Let $\vec{v}$ denote the velocity of a unit charge on C. Using the Lorentz force law, the tangential force per unit charge integrated along the circuit C is

$$\mathcal{E} = \oint_C d\vec{l}\cdot(\vec{\mathbf{E}} + \vec{v}\times\vec{\mathbf{B}}).$$

Note that if the circuit C is static in the space, then $\vec{v}$ is always pointed along the direction of the circuit, hence $d\vec{l}\cdot(\vec{v}\times\vec{\mathbf{B}}) = 0$. In this case, we have

$$\mathcal{E} = \oint_C d\vec{l}\cdot\vec{\mathbf{E}} \qquad \text{if } C \text{ is at rest.}$$

In other words, the second term in $\mathcal{E}$ comes from the effect of moving circuit. The precise relationship is described by the "**Flux Rule**" that we now derive.

Assume C bounds a surface S. The magnetic flux through S is defined to be

$$\Phi = \int_S d\vec{\sigma} \cdot \vec{\mathbf{B}}.$$

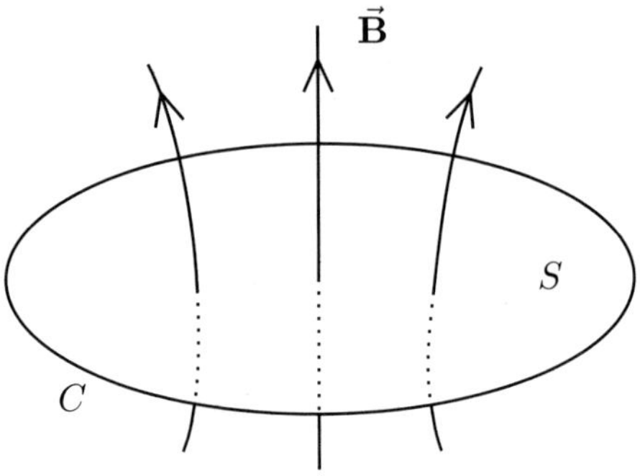

Let us now consider the change of Φ with respect to time. Assume the circuit C is moving with velocity $\vec{v}_C(\vec{r})$ at each point $\vec{r}$ of C. Let us write

$$\Phi = \int_S *_3\mathbb{B}$$

where $\mathbb{B} = \mathbf{B}_x dx + \mathbf{B}_y dy + \mathbf{B}_z dz$. Then

$$\frac{d}{dt}\Phi = \int_S \frac{\partial}{\partial t}(*_3\mathbb{B}) + \int_C \iota_{\vec{v}_C}(*_3\mathbb{B}).$$

Here $\iota_{\vec{v}_C}$ is the contraction with respect to the vector field $\vec{v}_C$. Expanding the notation, we find (Exercise)

$$\int_C \iota_{\vec{v}_C}(*_3\mathbb{B}) = -\int_C d\vec{l} \cdot (\vec{v}_C \times \vec{\mathbf{B}}).$$

This can be also understood from the picture: the total charge of the flux is computed by the "volume" formed by $d\vec{l}, \vec{v}_C, \vec{\mathbf{B}}$. We know that the "volume" is computed by the determinant, hence the expression $d\vec{l} \cdot (\vec{v}_C \times \vec{\mathbf{B}})$.

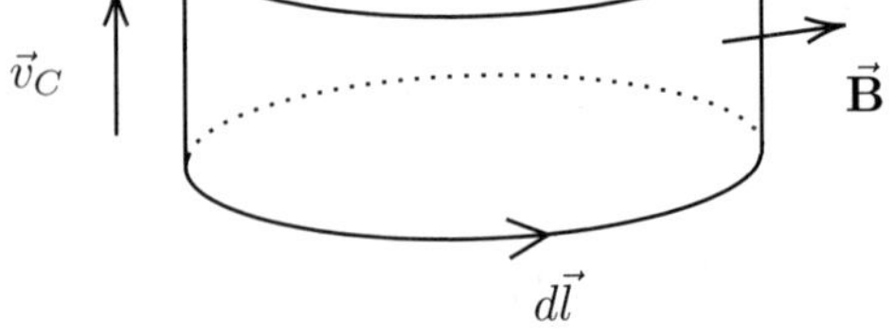

It follows that

$$\begin{aligned}
\frac{d\Phi}{dt} &= \int_S d\vec{\sigma} \cdot \frac{\partial \vec{\mathbf{B}}}{\partial t} - \int_C d\vec{l} \cdot (\vec{v}_C \times \vec{\mathbf{B}}) \\
&\overset{\text{Faraday's}}{\underset{\text{law}}{=\!=\!=}} - \int_S d\vec{\sigma} \cdot (\nabla \times \vec{\mathbf{E}}) - \int_C d\vec{l} \cdot (\vec{v}_C \times \vec{\mathbf{B}}) \\
&= - \int_C d\vec{l} \cdot \vec{\mathbf{E}} - \int_C d\vec{l} \cdot (\vec{v}_C \times \vec{\mathbf{B}}).
\end{aligned}$$

Now the velocity $\vec{v}$ differs from $\vec{v}_C$ by a tangential vector along C, hence

$$\int_C d\vec{l} \cdot (\vec{v}_C \times \vec{\mathbf{B}}) = \int_C d\vec{l} \cdot (\vec{v} \times \vec{\mathbf{B}}).$$

We have now arrived at

$$\textbf{Flux Rule}: \qquad \mathcal{E} = -\frac{d\Phi}{dt}$$

the change of the magnetic flux is the emf.

To understand the minus sign, imagine that the emf will produce a current which itself will induce a magnetic field. The minus sign says that the induced magnetic field will always be in the direction that opposes the change. This is called **Lenz's Law**. As a result, the Faraday law of induction is similar to the "inertial" phenomenon in mechanics. A conducting circuit prefers to maintain a constant flux through it. The change of the flux will result in a responding current in such a direction to oppose the change.

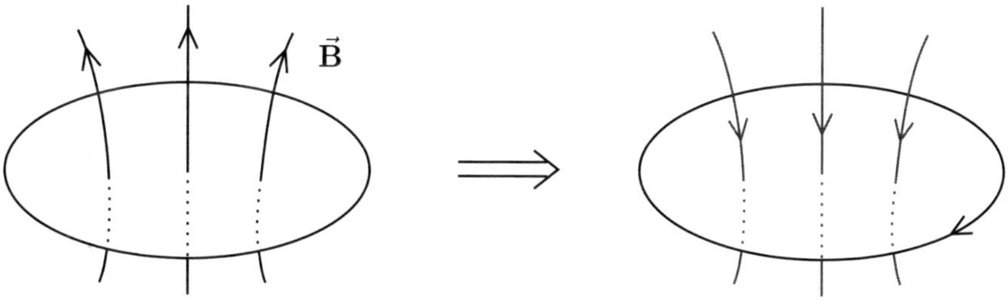

assume $\vec{\mathbf{B}}$ increases — emf induced current and magnetic field

Example 3.2.1 (Jumping Ring). Place a metal ring on top of a solenoidal coil around an iron core. Let us now switch on the current, which will produce a magnetic field. This will induce a current in the ring in the opposite direction. Since opposite current repel, the ring will jump out.

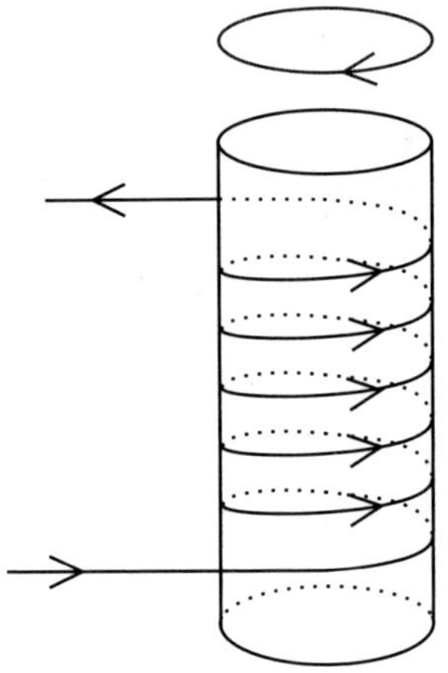

3.2.2 Mutual Inductance

Consider two loops C_1, C_2 of wire at rest. If we run a steady current I_1 around C_1, it will produce a magnetic field $\vec{\mathbf{B}}_1$ and hence a magnetic flux Φ_2 through the loop C_2. The flux Φ_2 is proportional to the current I_1

$$\Phi_2 = M_{21} I_1$$

where the constant M_{21} is called the mutual inductance of C_1 and C_2.

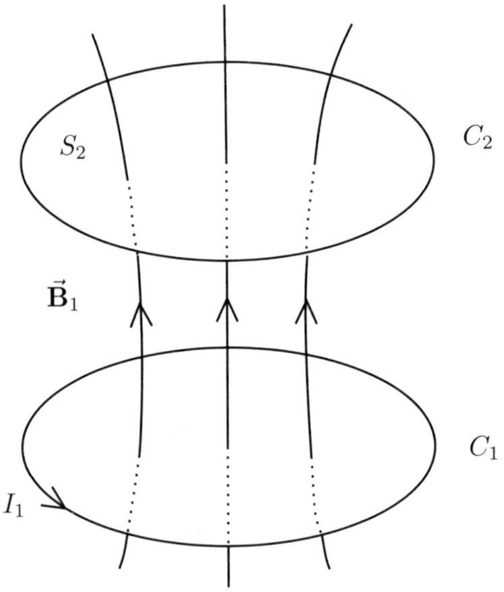

In Coulomb gauge, the vector potential $\vec{\mathbf{A}}_1$ produced by I_1 is

$$\vec{\mathbf{A}}_1(\vec{r}) = \frac{\mu_0 I_1}{4\pi} \oint_{C_1} \frac{d\vec{r}\,'}{|\vec{r} - \vec{r}\,'|}.$$

Therefore

$$\Phi_2 = \int_{S_2} d\vec{\sigma} \cdot \vec{\mathbf{B}}_1 = \int_{S_2} d\vec{\sigma} \cdot (\nabla \times \vec{\mathbf{A}}_1) = \oint_{C_2} d\vec{r} \cdot \vec{\mathbf{A}}_1(\vec{r}) = \frac{\mu_0 I_1}{4\pi} \oint_{C_2} \oint_{C_1} \frac{d\vec{r} \cdot d\vec{r}'}{|\vec{r} - \vec{r}'|}.$$

It follows that

$$M_{21} = \frac{\mu_0}{4\pi} \oint_{C_2} \oint_{C_1} \frac{d\vec{r} \cdot d\vec{r}'}{|\vec{r} - \vec{r}'|}$$

which is a purely geometric quantity. This expression is invariant under the switch of C_1 and C_2, hence

$$M_{21} = M_{12} \overset{\text{call}}{=\!=\!=} M.$$

As a consequence, if we vary the current in C_1, Faraday's law of induction says this will vary the flow through C_2 and produce a current in C_2. Isn't this amazing!

3.2.3 Self Inductance

A changing current also induces an emf in the source loop itself. The flux is proportional to the current

$$\Phi = LI$$

where L is called the self induction of the loop. The induced emf is then given by

$$\mathcal{E} = -L\frac{dI}{dt}.$$

Example 3.2.2 (Inductance of the Solenoid)**.** Consider a solenoid wrapped by a wire which carries current I and winds n times per unit length. Assume the solenoid has length l and cross-sectional area A, and $l >> \sqrt{A}$. This is approximately the infinite cylinder example with horizontal surface current density NI as we discussed before. The magnetic field inside the solenoid is

$$\vec{\mathbf{B}} = \mu_0 N I \hat{z}.$$

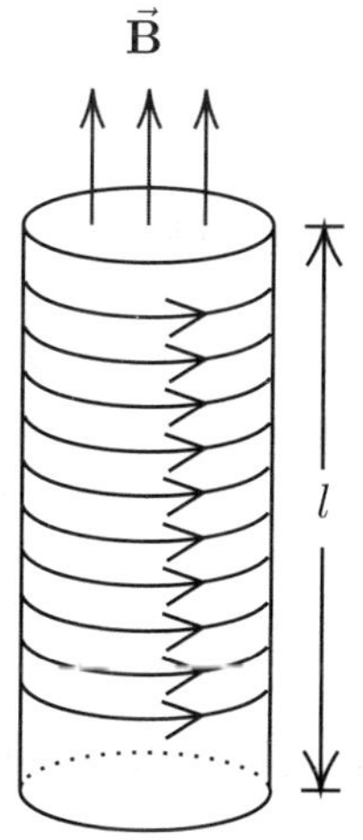

The wire winds Nl times, hence the total flux is

$$\Phi = (\mu_0 NIA)Nl = \mu_0 IN^2 Al = \mu_0 IN^2 V$$

where $V = Al$ is the volume of the solenoid. The self-inductance is therefore $L = \mu_0 N^2 V$.

3.2.4 Magnetostatic Energy

Let us try to increase the current of a circuit C to achieve the current I. The work done in this process can be viewed as the energy stored in the circuit. At time t, the induced emf is

$$\mathcal{E} = -L\frac{dI}{dt}.$$

Therefore the rate of electrical work by induced forces is

$$\frac{dW}{dt} = \mathcal{E}I = -LI\frac{dI}{dt},$$

so

$$W = -\frac{1}{2}LI^2.$$

The total energy stored is $U = -W = \frac{1}{2}LI^2$. It is illustrating to compare this with the mechanical case

Particle	Circuit
mv (momentum)	LI
$\frac{1}{2}mv^2$ (kinetic energy)	$\frac{1}{2}LI^2$

To see this is the correct magnetostatic energy, we can write it as

$$U = \frac{1}{2}I(LI) = \frac{1}{2}I\Phi = \frac{1}{2}I\int_S d\vec{\sigma}\cdot\vec{\mathbf{B}} = \frac{1}{2}I\oint_C d\vec{l}\cdot\vec{\mathbf{A}}.$$

Let us rewrite this into the form of current density by replacing $Id\vec{l} \to \vec{\mathbf{j}}\,d^3r$

$$U = \frac{1}{2}\int d^3r\,\vec{\mathbf{j}}\cdot\vec{\mathbf{A}}$$

In the magnetostatic case, this is

$$U = \frac{1}{2\mu_0}\int d^3r\,(\nabla\times\vec{\mathbf{B}})\cdot\vec{\mathbf{A}} = \frac{1}{2\mu_0}\int d^3r\,(\nabla\times\vec{\mathbf{A}})\cdot\vec{\mathbf{B}} = \frac{1}{2\mu_0}\int d^3r\,\vec{\mathbf{B}}\cdot\vec{\mathbf{B}}$$

which is the same electromagnetic energy (when $\vec{\mathbf{E}} = 0$) that we find previously in Section 3.1.2.

Example 3.2.3 (Solenoid)**.** We consider the solenoid as described in the previous example. The energy density of the solenoid is

$$\frac{1}{2\mu_0}\vec{\mathbf{B}}\cdot\vec{\mathbf{B}} = \frac{\mu_0 N^2 I^2}{2}.$$

The total stored energy is

$$U = \frac{\mu_0 N^2 I^2}{2} V.$$

Comparing with the formula $U = \frac{1}{2}LI^2$, we find

$$L = \mu_0 N^2 V$$

which coincides with the previous calculation of self inductance.

3.3 Electromagnetic Wave

We study Maxwell's equations in the absence of sources, which gives rise to solutions by electromagnetic waves propagating in spacetime. In particular, this leads to Maxwell's speculation of light as electromagnetic wave, and brings light, electricity and magnetism into the same fundamental phenomenon.

3.3.1 Wave Equation

Maxwell's Equations in the vacuum without sources are

$$\begin{cases} \nabla\cdot\vec{\mathbf{E}} = 0 \\ \nabla\cdot\vec{\mathbf{B}} = 0 \\ \nabla\times\vec{\mathbf{E}} = -\frac{\partial\vec{\mathbf{B}}}{\partial t} \\ \nabla\times\vec{\mathbf{B}} = \mu_0\varepsilon_0\frac{\partial\vec{\mathbf{E}}}{\partial t} \end{cases}$$

Here $\mu_0\varepsilon_0 = 1/c^2$. In terms of the electromagnetic 2-form

$$\begin{aligned} F &= \mathbb{E}\wedge dt + *_3\mathbb{B} = \mathbb{E}\wedge dt + *(\mathbb{B}\wedge cdt) \\ &= (\mathbf{E}_x dx + \mathbf{E}_y dy + \mathbf{E}_z dz)\wedge dt + (\mathbf{B}_x dy\wedge dz + \mathbf{B}_y dz\wedge dx + \mathbf{B}_z dx\wedge dy), \end{aligned}$$

the above equations have the geometric form

$$\begin{cases} dF = 0 \\ d^*F = 0 \end{cases}$$

Here $d^* = *d*$ is the adjoint of d in $\mathbb{R}^{3,1}$. Let

$$\Box = \frac{1}{c^2}\frac{\partial^2}{\partial t^2} - \nabla^2 = \frac{1}{c^2}\frac{\partial^2}{\partial t^2} - \frac{\partial^2}{\partial x^2} - \frac{\partial^2}{\partial y^2} - \frac{\partial^2}{\partial z^2}$$

be the d'Alembert operator. Using the relation

$$dd^* + d^*d = -\Box,$$

we find that F satisfies

$$\Box F = 0.$$

In components, this is

$$\begin{cases} \frac{1}{c^2}\frac{\partial^2 \vec{\mathbf{E}}}{\partial t^2} - \nabla^2 \vec{\mathbf{E}} = 0 \\ \frac{1}{c^2}\frac{\partial^2 \vec{\mathbf{B}}}{\partial t^2} - \nabla^2 \vec{\mathbf{B}} = 0 \end{cases}$$

which are the vector valued wave equations travelling at the speed of light c.

Since Maxwell's Equations in the vacuum are linear, it turns out to be convenient to express solutions via complex valued functions while electromagnetic fields are obtained by taking their real part. We will assume this implicitly and will not distinguish real and complex valued solutions when it is clear from the context.

3.3.2 Plane Waves

Let $\vec{k} = (k_x, k_y, k_z) \in \mathbb{R}^3$ be a constant vector in $\mathbb{R}^3$. A standard plane wave solving the scalar wave equation

$$\Box \varphi = 0$$

propagating along $\vec{k}$ is given by the form

$$\varphi = e^{i(\vec{k}\cdot\vec{r} - \omega t)}.$$

Plug into the wave equation, we find

$$\Box \varphi = 0 \qquad \Longleftrightarrow \qquad \omega^2 = c^2 k^2 \quad \text{where } k = |\vec{k}|.$$

The two cases $\omega = \pm ck$ corresponding to two waves propagating in opposite directions along $\vec{k}$. We will take

$$\omega = ck$$

in the following discussion. The other case can be equivalently described by the propagating vector $-\vec{k}$.

To see the precise meaning of $\vec{k}$ and ω, let us first consider standing at a fixed point $\vec{r}$ in the space. Then the wave $e^{i(\vec{k}\cdot\vec{r}-\omega t)}$ at the point $\vec{r}$ will oscillate $\frac{\omega}{2\pi}$ times per unit time. Similarly, if we look at the wave at an instant time t, we find that the wave will oscillate $\frac{k}{2\pi}$ times per unit length along the direction $\vec{k}$.

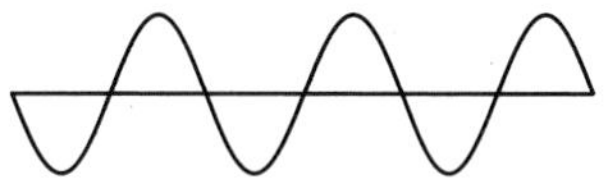

$\frac{\omega}{2\pi}$ times per unit time $\qquad$ $\frac{k}{2\pi}$ times per unit length

So $\frac{\omega}{2\pi}/\frac{k}{2\pi} = \omega/k$ is the velocity of the wave.

Remark 3.3.1. The ratio ω/k is called the "**phase velocity**" of the wave. For electromagnetic waves in the vacuum, the phase velocity is always the speed of light c for waves of any frequency: $\omega/k = c$. This is called "dispersionless" waves. If we consider waves in dispersive matter, we will find the phase velocity depending on frequency

$$\omega(k) = v(k) \cdot k.$$

Such equation is generally called the **dispersion relation**.

Let us now describe an electromagnetic plane wave given by

$$\begin{cases} \vec{\mathbf{E}} = \vec{\mathbf{E}}_0 e^{i(\vec{k}\cdot\vec{r}-\omega t)} \\ \vec{\mathbf{B}} = \vec{\mathbf{B}}_0 e^{i(\vec{k}\cdot\vec{r}-\omega t)} \end{cases} \qquad \text{where } \omega = ck.$$

Here $\vec{\mathbf{E}}_0, \vec{\mathbf{B}}_0$ are complex valued constant vectors. Physical solutions are their real parts as remarked above. Since $\vec{\mathbf{E}}_0, \vec{\mathbf{B}}_0$ are constants, they solve the wave equation

$$\Box\vec{\mathbf{E}} = \Box\vec{\mathbf{B}} = 0.$$

On the other hand, Maxwell's equations say more than wave equations and will put further constraints. The divergence free relations

$$\nabla \cdot \vec{\mathbf{E}} = 0, \qquad \nabla \cdot \vec{\mathbf{B}} = 0$$

is equivalent to

$$\vec{\mathbf{E}}_0 \cdot \vec{k} = 0, \qquad \vec{\mathbf{B}}_0 \cdot \vec{k} = 0.$$

This means that $\vec{\mathbf{E}}$ and $\vec{\mathbf{B}}$ are both perpendicular to the propagating direction $\vec{k}$. The curl relations

$$\nabla \times \vec{\mathbf{E}} = -\frac{\partial \vec{\mathbf{B}}}{\partial t}, \qquad \nabla \times \vec{\mathbf{B}} = \frac{1}{c^2}\frac{\partial \vec{\mathbf{E}}}{\partial t}$$

is equivalent to

$$\vec{k} \times \vec{\mathbf{E}}_0 = \omega \vec{\mathbf{B}}_0, \qquad \vec{k} \times \vec{\mathbf{B}}_0 = -\frac{\omega}{c^2}\vec{\mathbf{E}}_0.$$

This means that $\vec{\mathbf{E}}_0$ is also perpendicular to $\vec{\mathbf{B}}_0$ and

$$c\vec{\mathbf{B}}_0 = \hat{k} \times \vec{\mathbf{E}}_0 \qquad \text{where } \hat{k} = \frac{\vec{k}}{k}.$$

The above relations can be summarized by the geometric figure

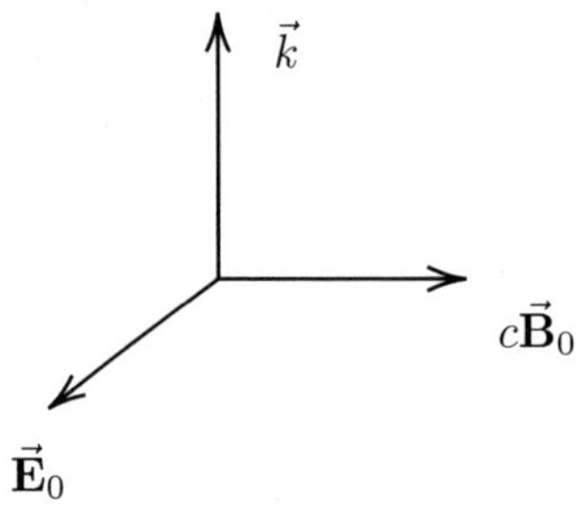

$$|\vec{\mathbf{E}}_0| = c|\vec{\mathbf{B}}_0|$$

3.3.3 Polarization

Let us consider electromagnetic waves in the propagating direction $\vec{k}$. Let us fix two unit vectors $\hat{e}_1, \hat{e}_2 \in \mathbb{R}^3$ such that

$$\hat{e}_1 \cdot \hat{e}_2 = 0, \qquad \hat{e}_1 \times \hat{e}_2 = \hat{k}.$$

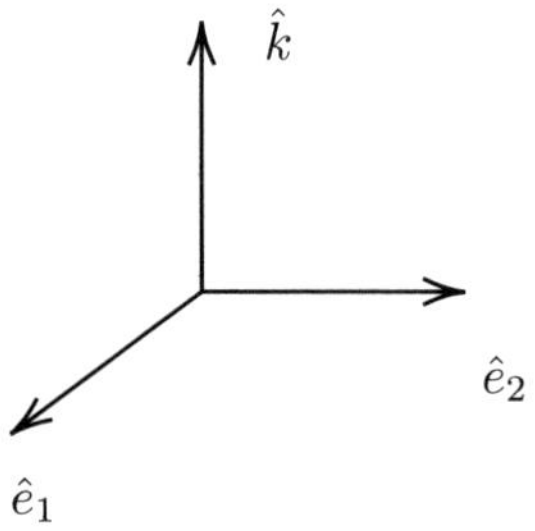

Let us express

$$\vec{\mathbf{E}}_0 = \mathbf{E}_1 \hat{e}_1 + \mathbf{E}_2 \hat{e}_2$$

where we emphasise that $\mathbf{E}_1, \mathbf{E}_2$ are complex numbers. Then

$$\vec{\mathbf{E}} = (\mathbf{E}_1 \hat{e}_1 + \mathbf{E}_2 \hat{e}_2)\, e^{i(\vec{k}\cdot\vec{r} - \omega t)}.$$

The magnetic field $\vec{\mathbf{B}}$ can be obtained from $\vec{\mathbf{E}}$ by

$$\vec{\mathbf{B}} = \frac{1}{c}\hat{k} \times \vec{\mathbf{E}} = \frac{1}{c}\left(-\mathbf{E}_2 \hat{e}_1 + \mathbf{E}_1 \hat{e}_2\right) e^{i(\vec{k}\cdot\vec{r} - \omega t)}.$$

Let us focus on the electric field. The physical field is given by the real part

$$\mathrm{Re}\left[(\mathbf{E}_1 \hat{e}_1 + \mathbf{E}_2 \hat{e}_2)\, e^{i(\vec{k}\cdot\vec{r} - \omega t)}\right].$$

Let us write

$$\mathbf{E}_1 = A_1 e^{i\theta_1}, \qquad \mathbf{E}_2 = A_2 e^{i\theta_2}, \qquad A_i \geq 0.$$

The behavior of the electric/magnetic field can be described via the following cases:

① Linear Polarization: $e^{i\theta_2} = \pm e^{i\theta_1}$ or $\theta_2 - \theta_1 \in \mathbb{Z}\pi$. In this case we have

$$\vec{\mathbf{E}} = \mathrm{Re}\left[\vec{u}e^{i(\vec{k}\cdot\vec{r}-\omega t+\theta_1)}\right] = \vec{u}\cos(\vec{k}\cdot\vec{r}-\omega t+\theta_1)$$

where $\vec{u}$ is the fixed vector $\vec{u} = A_1\hat{e}_1 \pm A_2\hat{e}_2$. The electromagnetic fields are oscillating along fixed directions. The wave at an instant time looks like

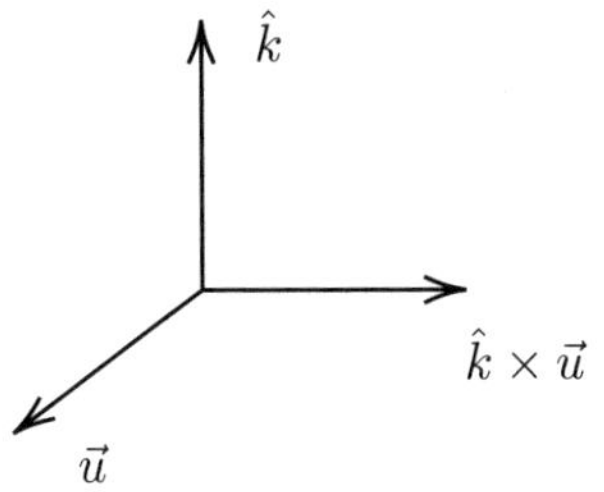

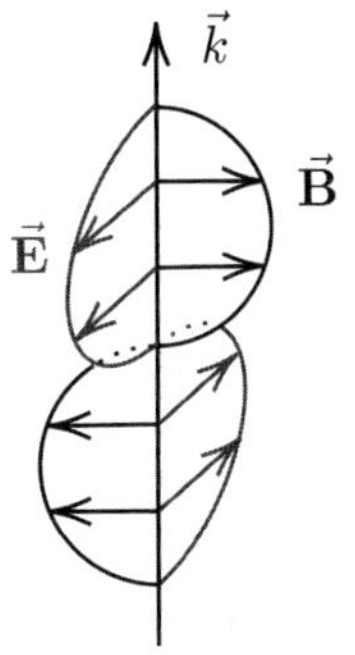

② Circular Polarization: $\mathbf{E}_2 = \pm i\mathbf{E}_1$. In this case we have

$$\begin{aligned}\vec{\mathbf{E}}_\pm &= \mathrm{Re}\left[(\hat{e}_1 \pm i\hat{e}_2)A_1e^{i(\vec{k}\cdot\vec{r}-\omega t+\theta_1)}\right]\\ &= A_1\left(\cos(\vec{k}\cdot\vec{r}-\omega t+\theta_1)\hat{e}_1 \mp \sin(\vec{k}\cdot\vec{r}-\omega t+\theta_1)\hat{e}_2\right).\end{aligned}$$

At a fixed position $\vec{r}$ in the space, the vector $\vec{\mathbf{E}}_\pm$ has a constant magnitude, but rotates in a circle at a frequency ω. $\vec{\mathbf{E}}_+$ is rotating counterclockwise, while $\vec{\mathbf{E}}_-$ is rotating clockwise, as shown below

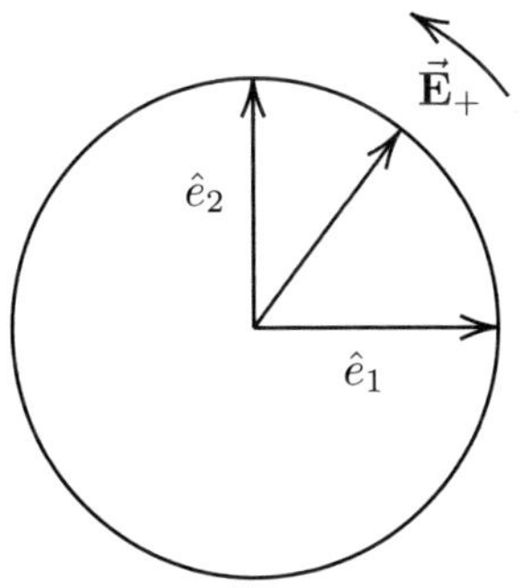

Left circular polarization

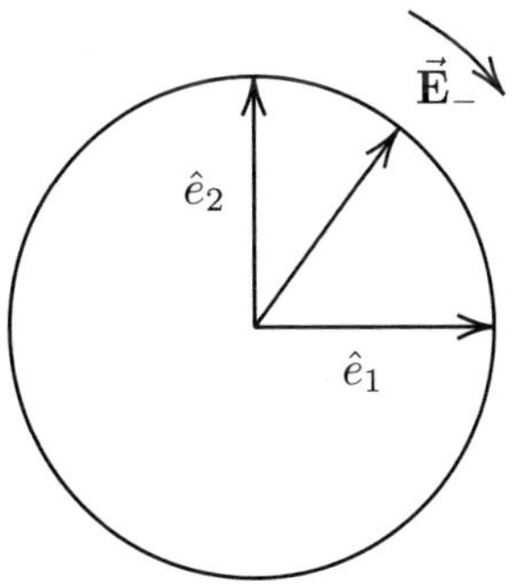

Right circular polarization

The wave at an instant time looks like (for the right circular polarization case)

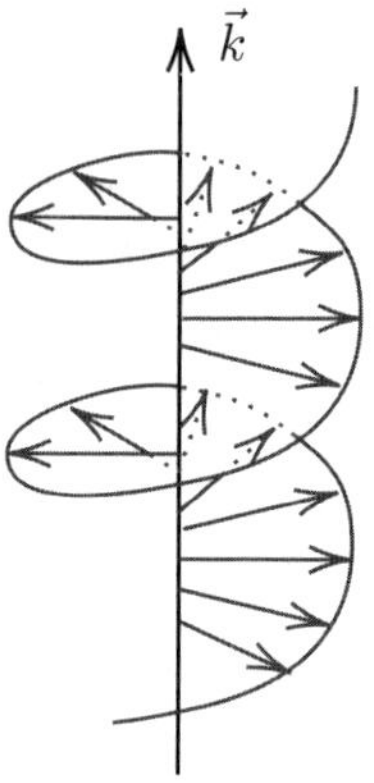

③ Elliptical Polarization: general $\mathbf{E}_1$ and $\mathbf{E}_2$.

$$\mathrm{Re}\left[(\mathbf{E}_1\hat{e}_1 + \mathbf{E}_2\hat{e}_2)e^{i(\vec{k}\cdot\vec{r}-\omega t)}\right] = A_1\cos(\vec{k}\cdot\vec{r}-\omega t+\theta_1)\hat{e}_1 + A_2\cos(\vec{k}\cdot\vec{r}-\omega t+\theta_2)\hat{e}_2.$$

It swaps out an ellipse at any point in the space.

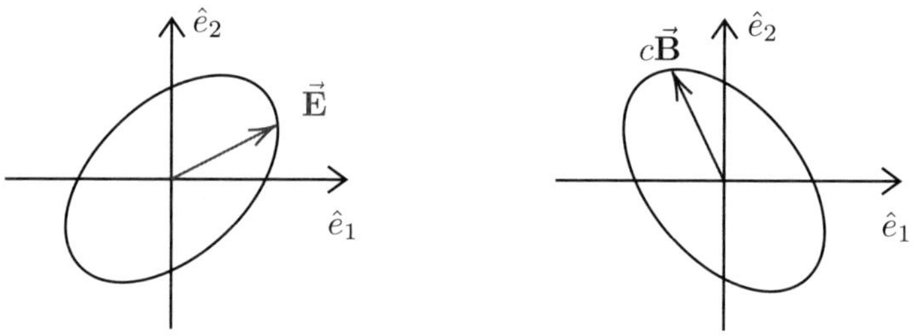

3.3.4 Wave Packets

In the above, we have focused on the case of a plane wave with fixed $\vec{k}$. Such a wave $e^{i(\vec{k}\cdot\vec{r}-\omega t)}$ spreads out the full space and can not exist in nature. In general, since the wave equation is linear, a localized wave can be obtained by superposing plane waves (like Fourier modes) with different wave vectors. A general electromagnetic wave packet is given by

$$\vec{\mathbf{E}} = \frac{1}{(2\pi)^3}\int d^3k\, \vec{\mathcal{E}}(\vec{k}\,)e^{i(\vec{k}\cdot\vec{r}-\omega t)}$$
$$\vec{\mathbf{B}} = \frac{1}{(2\pi)^3}\int d^3k\, \frac{1}{c}\left(\hat{k}\times\vec{\mathcal{E}}(\vec{k}\,)\right) e^{i(\vec{k}\cdot\vec{r}-\omega t)}$$

or more precisely their real part. The vector function $\vec{\mathcal{E}}(\vec{k}\,)$ plays the role of Fourier coefficients. In the vacuum which is dispersionless, we always have $\omega = ck$. In dispersive matters, we have a more complicated dispersion relation between ω and k.

Recall that the energy density of an electromagnetic field is

$$u = \frac{1}{2}\left(\varepsilon_0|\text{Re }\vec{\mathbf{E}}|^2 + \frac{1}{\mu_0}|\text{Re }\vec{\mathbf{B}}|^2\right) = \frac{\varepsilon_0}{2}\left(|\text{Re }\vec{\mathbf{E}}|^2 + |\text{Re }c\vec{\mathbf{B}}|^2\right)$$

and the total energy U is

$$U = \frac{\varepsilon_0}{2}\int d^3r\,\left(|\text{Re }\vec{\mathbf{E}}|^2 + |\text{Re }c\vec{\mathbf{B}}|^2\right).$$

In the vacuum case ($\omega = ck$), we calculate

$$\begin{aligned}|\text{Re }\vec{\mathbf{E}}|^2 &= \frac{1}{4}(\vec{\mathbf{E}}\cdot\vec{\mathbf{E}} + \vec{\mathbf{E}}^*\cdot\vec{\mathbf{E}}^* + 2\vec{\mathbf{E}}\cdot\vec{\mathbf{E}}^*) = \frac{1}{4}(\vec{\mathbf{E}}\cdot\vec{\mathbf{E}} + \vec{\mathbf{E}}\cdot\vec{\mathbf{E}}^*) + \text{c.c.}\\ &= \frac{1}{4}\frac{1}{(2\pi)^3}\int d^3k\,\frac{1}{(2\pi)^3}\int d^3\vec{k}'\,\Big(\vec{\mathcal{E}}(\vec{k})\cdot\vec{\mathcal{E}}(\vec{k}')e^{i(\vec{k}+\vec{k}')\cdot\vec{r}-ic(k+k')t}\\ &\quad + \vec{\mathcal{E}}(\vec{k})\cdot\vec{\mathcal{E}}^*(\vec{k}')e^{i(\vec{k}-\vec{k}')\cdot\vec{r}-ic(k-k')t}\Big) + \text{c.c.}\end{aligned}$$

Here c.c. means the complex conjugate of the expression before it. Using the identity

$$\frac{1}{(2\pi)^3}\int d^3r\,e^{i\vec{p}\cdot\vec{r}} = \delta(\vec{p}),$$

we find

$$\int d^3r\,|\text{Re }\vec{\mathbf{E}}|^2 = \frac{1}{4}\frac{1}{(2\pi)^3}\int d^3k\,\left(\vec{\mathcal{E}}(\vec{k})\cdot\vec{\mathcal{E}}^*(\vec{k}) + \vec{\mathcal{E}}(\vec{k})\cdot\vec{\mathcal{E}}(-\vec{k})e^{-2ickt}\right) + c.c.$$

Similarly, the contribution from the magnetic field is

$$\begin{aligned}&\int d^3r\,|\text{Re }c\vec{\mathbf{B}}|^2\\ &= \frac{1}{4}\frac{1}{(2\pi)^3}\int d^3k\,\left[\left(\hat{k}\times\vec{\mathcal{E}}(\vec{k})\right)\cdot\left(\hat{k}\times\vec{\mathcal{E}}^*(\vec{k})\right) + \left(\hat{k}\times\vec{\mathcal{E}}(\vec{k})\right)\cdot\left(-\hat{k}\times\vec{\mathcal{E}}(-\vec{k})\right)e^{-2ickt}\right]\\ &\quad + c.c.\\ &= \frac{1}{4}\frac{1}{(2\pi)^3}\int d^3k\,\left(\vec{\mathcal{E}}(\vec{k})\cdot\vec{\mathcal{E}}^*(\vec{k}) - \vec{\mathcal{E}}(\vec{k})\cdot\vec{\mathcal{E}}(-\vec{k})e^{-2ickt}\right) + c.c.\end{aligned}$$

Adding the above two contributions, we find the following total energy formula for electromagnetic wave packet in a region of vacuum

$$U = \frac{\varepsilon_0}{2}\frac{1}{(2\pi)^3}\int d^3k\,|\vec{\mathcal{E}}(\vec{k})|^2.$$

3.4 Green's Functions

3.4.1 Wave Equation with Source

We now consider solving the full Maxwell's equations with specified distributions of charge and current:

$$\begin{cases} \nabla \cdot \vec{\mathbf{E}} = \rho/\varepsilon_0 \\ \nabla \cdot \vec{\mathbf{B}} = 0 \\ \nabla \times \vec{\mathbf{E}} = -\frac{\partial \vec{\mathbf{B}}}{\partial t} \\ \nabla \times \vec{\mathbf{B}} = \mu_0 \left(\vec{\mathbf{j}} + \varepsilon_0 \frac{\partial \vec{\mathbf{E}}}{\partial t}\right) \end{cases}$$

In terms of the electromagnetic 2-form F and the charge current 1-form J

$$F = \mathbb{E} \wedge dt + *_3 \mathbb{B}, \qquad J = \rho/\varepsilon_0 \, dt - \mu_0 \mathbb{j},$$

Maxwell's equations take the geometric form

$$\begin{cases} dF = 0 \\ d^* F = J \end{cases}$$

Here $d^* = *d*$ is the adjoint of d in $\mathbb{R}^{3,1}$. Recall

$$d^2 = 0, \qquad (d^*)^2 = 0, \qquad dd^* + d^* d = -\Box.$$

The consistency condition for $d^* F = J$ requires

$$d^* J = 0.$$

Since $d^* J = -\frac{1}{c^2 \varepsilon_0} \partial_t \rho - \mu_0 \nabla \cdot \vec{\mathbf{j}} = -\mu_0 (\partial_t \rho + \nabla \cdot \vec{\mathbf{j}})$, the consistency condition is precisely the charge conservation as we discussed in Section 1.3

$$\partial_t \rho + \nabla \cdot \vec{\mathbf{j}} = 0.$$

The equation $dF = 0$ can be solved by introducing the potential 1-form A on $\mathbb{R}^{3,1}$ such that

$$F = dA.$$

Expressing in components,

$$A = -\phi dt + \mathbb{A}$$

where ϕ is the scalar potential and $\mathbb{A} = A_x dx + A_y dy + A_z dz$ is the vector potential. Now in general, ϕ and $\mathbb{A}$ depends on the position $\vec{r}$ and time t. The equation $F = dA$ reads in components by

$$\begin{cases} \vec{\mathbf{E}} = -\nabla \phi - \partial_t \vec{\mathbf{A}} \\ \vec{\mathbf{B}} = \nabla \times \vec{\mathbf{A}} \end{cases}$$

The choice of A is not unique, and we have the following gauge transformation leaving F invariant

$$A \longmapsto A + d\chi$$

for any function χ on $\mathbb{R}^{3,1}$. We need a gauge fixing condition to specify the solution of A as discussed in Section 1.10. We will focus on the Lorenz gauge here:

$$\text{Lorenz gauge: } d^*A = 0 \qquad \Longleftrightarrow \qquad \frac{1}{c^2}\frac{\partial\phi}{\partial t} + \nabla\cdot\vec{\mathbf{A}} = 0.$$

In terms of the potential 1-form A, the other half of Maxwell's Equations $d^*F = J$ becomes

$$d^*dA = J.$$

In Lorenz gauge,

$$\begin{cases} d^*dA = J & \text{Maxwell's Equations} \\ d^*A = 0 & \text{Lorenz gauge} \end{cases}$$

from which we find

$$\Box A = -J.$$

In components, this becomes

$$\begin{cases} \Box\phi = \rho/\varepsilon_0 \\ \Box\vec{\mathbf{A}} = \mu_0\vec{\mathbf{j}} \end{cases}$$

These are **inhomogeneous wave equations** with source terms.

3.4.2 Green's Function for the Wave Equation

To solve the inhomogeneous wave equation, we follow the same strategy as before by constructing the inverse "$\frac{1}{\Box}$", which is called the Green's function for the wave equation.

By definition, the Green's function $G(\vec{r},t;\vec{r}\,',t')$ for the wave equation is the wave produced at $(\vec{r}\,',t')$ by a unit point source at $(\vec{r},t)$. It satisfies

$$\Box' G(\vec{r},t;\vec{r}\,',t') = \delta(\vec{r}-\vec{r}\,')\delta(t-t').$$

Here $\Box' = \frac{1}{c^2}\frac{\partial^2}{\partial t'^2} - \nabla^{2\prime}$ acts on the variable $(\vec{r}\,',t')$.

Let us first explain how to solve inhomogeneous wave equation using the Green's function, and come back to the construction of the Green's function later. Consider a function $\varphi(\vec{r},t)$ satisfying the inhomogeneous wave equation

$$\Box\varphi = f.$$

To find the expression of φ in terms of f, we apply the Green's function in the region $V \in \mathbb{R}^3$ and time interval $[t_1,t_2]$:

$$\begin{aligned} \varphi(\vec{r},t) &= \int_{t_1}^{t_2} dt' \int_V d^3r'\, \varphi(\vec{r}\,',t')\delta(\vec{r}-\vec{r}\,')\delta(t-t') \\ &= \int_{t_1}^{t_2} dt' \int_V d^3r'\, \varphi(\vec{r}\,',t')\left(\frac{1}{c^2}\frac{\partial^2}{\partial t'^2} - \nabla^{2\prime}\right) G(\vec{r},t;\vec{r}\,',t'). \end{aligned}$$

For the first term, we use integration by part and find

$$\int_{t_1}^{t_2} dt' \, \varphi \frac{\partial^2}{\partial t'^2} G = \int_{t_1}^{t_2} dt' \left(\frac{\partial^2}{\partial t'^2} \varphi \right) G + \left(\varphi \frac{\partial G}{\partial t'} - G \frac{\partial \varphi}{\partial t'} \right) \Big|_{t_1}^{t_2}.$$

For the second term, we have as before

$$\int_V d^3 r' \, \varphi \nabla^{2\prime} G = \int_V d^3 r' \, (\nabla^{2\prime} \varphi) \, G + \int_{\partial V} (\varphi \partial'_{\hat{n}} G - G \partial'_{\hat{n}} \varphi)$$

where $\hat{n}$ is the unit normal vector on the surface ∂V.

Combining the above two equations and using $\Box \varphi = f$, we find

$$\begin{aligned} \varphi(\vec{r}, t) = & \int_{t_1}^{t_2} dt' \int_V d^3 r' \, G(\vec{r}, t; \vec{r}\,', t') f(\vec{r}\,', t') \\ & + \int_{t_1}^{t_2} dt' \int_{\partial V} \left[G(\vec{r}, t; \vec{r}\,', t') \partial'_{\hat{n}} \varphi(\vec{r}\,', t') - \varphi(\vec{r}\,', t') \partial'_{\hat{n}} G(\vec{r}, t; \vec{r}\,', t') \right] \\ & + \frac{1}{c^2} \int_V d^3 r' \left[\varphi(\vec{r}\,', t') \partial_{t'} G(\vec{r}, t; \vec{r}\,', t') - G(\vec{r}, t; \vec{r}\,', t') \partial_{t'} \varphi(\vec{r}\,', t') \right] \Big|_{t'=t_1}^{t'=t_2}. \end{aligned}$$

The first term is the superposition of unit point source by the source function f. The second term is the boundary term in the spatial variables. The third term is the boundary term in the time variable.

Let us now construct the Green's function in $\mathbb{R}^{3,1}$. By symmetry considerations, we seek a Green's function of the form

$$G(\vec{r}, t; \vec{r}\,', t') = G(\vec{r} - \vec{r}\,', t - t')$$

and the function $G(\vec{r}, t)$ satisfying

$$\left(\frac{1}{c^2} \frac{\partial^2}{\partial t^2} - \nabla^2 \right) G(\vec{r}, t) = \delta(\vec{r}) \delta(t).$$

Furthermore, it is natural to assume $G(\vec{r}, t)$ is spherically symmetric, and hence $G(\vec{r}, t) = G(r, t)$ only depends on the radius $r = |\vec{r}|$ and time t. Then

$$\nabla^2 G = \frac{1}{r^2} \frac{\partial}{\partial r} \left(r^2 \frac{\partial G}{\partial r} \right) = \frac{\partial^2 G}{\partial r^2} + \frac{2}{r} \frac{\partial G}{\partial r} = \frac{1}{r} \frac{\partial^2}{\partial r^2} (rG).$$

We are reduced to solve

$$\frac{1}{r} \left(\frac{1}{c^2} \frac{\partial^2}{\partial t^2} - \frac{\partial^2}{\partial r^2} \right) (rG) = \delta(\vec{r}) \delta(t).$$

At $r > 0$, we have the one dimensional wave equation

$$\left(\frac{1}{c^2} \frac{\partial^2}{\partial t^2} - \frac{\partial^2}{\partial r^2} \right) (rG) = \left(\frac{1}{c} \frac{\partial}{\partial t} - \frac{\partial}{\partial r} \right) \left(\frac{1}{c} \frac{\partial}{\partial t} + \frac{\partial}{\partial r} \right) (rG) = 0$$

which has two independent solutions of the form

$$G^{(\mp)}(r,t) = \frac{1}{r} g_{\pm}(t \pm r/c)$$

for functions $g_{\pm}$ to be determined from the behavior at $r = 0$. Using

$$\nabla^2 \left(\frac{1}{r}\right) = -4\pi\delta(\vec{r}),$$

we have the following singular behavior at $r = 0$

$$\left(\frac{1}{c^2}\frac{\partial^2}{\partial t^2} - \nabla^2\right) G^{(\mp)}(r,t) = 4\pi\delta(\vec{r}) g_{\pm}(t \pm r/c).$$

Comparing with the equation for Green's function, we find

$$g_{\pm}(u) = \frac{1}{4\pi}\delta(u).$$

We conclude that

$$G^{(\mp)}(\vec{r},t) = \frac{1}{4\pi r}\delta(t \pm r/c).$$

Therefore we find two solutions of Green's function

$$G^{(\mp)}(\vec{r},t;\vec{r}\,',t') = \frac{1}{4\pi|\vec{r}-\vec{r}\,'|}\delta(t - t' \pm |\vec{r}-\vec{r}\,'|/c).$$

3.4.3 Retarded and Advanced Solutions

Let us plug these Green's functions into the above expression of φ satisfying the inhomogeneous wave equation

$$\Box\varphi = f.$$

We assume a spatially localized source and such that the spatial boundary term is absent. Thus

$$\begin{aligned}\varphi(\vec{r},t) = &\int_{t_1}^{t_2} dt' \int_V d^3r'\, G(\vec{r},t;\vec{r}\,',t') f(\vec{r}\,',t') \\ &+ \frac{1}{c^2}\int_V d^3r' \left[\varphi(\vec{r}\,',t')\partial_{t'}G(\vec{r},t;\vec{r}\,',t') - G(\vec{r},t;\vec{r}\,',t')\partial_{t'}\varphi(\vec{r}\,',t')\right]\Bigg|_{t'=t_1}^{t'=t_2}.\end{aligned}$$

There are two cases:

① Retarded Solution: This is to use

$$G^{(+)}(\vec{r},t;\vec{r}\,',t') = \frac{1}{4\pi|\vec{r}-\vec{r}\,'|}\delta(t - t' - |\vec{r}-\vec{r}\,'|/c).$$

Note that in the expression for $\varphi(\vec{r},t)$, we have

$$t_1 \le t \le t_2$$

so $G^{(+)}$ is only nonzero at the $t' = t_1$ boundary. The retarded solution is therefore given by

$$\varphi_{\text{ret}}(\vec{r},t) = \frac{1}{4\pi}\int d^3r' \, \frac{f(\vec{r}', t - |\vec{r}-\vec{r}'|/c)}{|\vec{r}-\vec{r}'|} + \varphi_{\text{in}}(\vec{r},t)$$

where $\varphi_{\text{in}}(\vec{r},t)$ collects the boundary term at $t' = t_1$.

The term involving the source f has the following interpretation: $\varphi_{\text{ret}}(\vec{r},t)$ is contributed from source located at $\vec{r}'$ at time $t - |\vec{r}-\vec{r}'|/c$. Such time delay reflects the propagating wave signaled at the speed of light c.

Assume t_1 is at the time before the source appears. Then at $t \leq t_1$, the source integral does not contribute and we have

$$\varphi_{\text{ret}}(\vec{r},t) = \varphi_{\text{in}}(\vec{r},t), \qquad t \leq t_1.$$

So physically, $\varphi_{\text{in}}(\vec{r},t)$ can be viewed as an "incoming wave" solution of the homogeneous wave equation at the initial time t_1 before the effect of the source. This is also consistent with the fact that a general solution of the inhomogeneous wave equation is given by the sum of a special solution (here is the source term) and a solution of the homogeneous wave equation (here is the incoming wave).

② Advanced Solution: This is to use

$$G^{(-)}(\vec{r},t;\vec{r}',t') = \frac{1}{4\pi|\vec{r}-\vec{r}'|}\delta(t - t' + |\vec{r}-\vec{r}'|/c).$$

The discussion is similar and we find the advanced solution

$$\varphi_{\text{adv}}(\vec{r},t) = \frac{1}{4\pi}\int d^3r' \, \frac{f(\vec{r}', t + |\vec{r}-\vec{r}'|/c)}{|\vec{r}-\vec{r}'|} + \varphi_{\text{out}}(\vec{r},t).$$

The advanced solution at $(\vec{r},t)$ takes the effect of the source located at $\vec{r}$ at a future time $t + |\vec{r}-\vec{r}'|/c$. And φ_{out} is the "outgoing wave" describing the situation at time t_2 after the effect of the source.

Mathematically, both the retarded and the advanced waves are solutions. Physically, we will pick the retarded one which reflects the causal structure.

Solving Maxwell's Equations

We now use the retarded Green's function to write down a solution of the inhomogeneous wave equations

$$\begin{cases} \Box\phi = \rho/\varepsilon_0 \\ \Box\vec{\mathbf{A}} = \mu_0\vec{\mathbf{j}} \end{cases}$$

It reads

$$\begin{cases} \phi(\vec{r},t) = \dfrac{1}{4\pi\varepsilon_0}\displaystyle\int d^3r' \, \frac{\rho(\vec{r}', t - |\vec{r}-\vec{r}'|/c)}{|\vec{r}-\vec{r}'|} \\ \vec{\mathbf{A}}(\vec{r},t) = \dfrac{\mu_0}{4\pi}\displaystyle\int d^3r' \, \frac{\vec{\mathbf{j}}(\vec{r}', t - |\vec{r}-\vec{r}'|/c)}{|\vec{r}-\vec{r}'|} \end{cases}$$

Let us check the Lorenz gauge condition

$$\frac{1}{c^2}\frac{\partial}{\partial t}\phi + \nabla \cdot \vec{\mathbf{A}} = 0.$$

Let us write the above formula via Green's function

$$\begin{cases} \phi(\vec{r},t) = \dfrac{1}{\varepsilon_0}\displaystyle\int d^3r'\, dt'\, G^{(+)}(\vec{r},t;\vec{r}\,',t')\rho(\vec{r}\,',t') \\ \vec{\mathbf{A}}(\vec{r},t) = \mu_0 \displaystyle\int d^3r'\, dt'\, G^{(+)}(\vec{r},t;\vec{r}\,',t')\vec{\mathbf{j}}(\vec{r}\,',t') \end{cases}$$

and using $\varepsilon_0\mu_0 = 1/c^2$, we find

$$\begin{aligned} &\frac{1}{c^2}\frac{\partial}{\partial t}\phi + \nabla \cdot \vec{\mathbf{A}} \\ &= \mu_0 \int d^3r'\, dt' \left[\partial_t G^{(+)}(\vec{r},t;\vec{r}\,',t')\rho(\vec{r}\,',t') + \nabla \cdot G^{(+)}(\vec{r},t;\vec{r}\,',t')\vec{\mathbf{j}}(\vec{r}\,',t')\right] \\ &= -\mu_0 \int d^3r'\, dt' \left[\partial'_t G^{(+)}(\vec{r},t;\vec{r}\,',t')\rho(\vec{r}\,',t') + \nabla' \cdot G^{(+)}(\vec{r},t;\vec{r}\,',t')\vec{\mathbf{j}}(\vec{r}\,',t')\right] \\ &= \mu_0 \int d^3r'\, dt'\, G^{(+)}(\vec{r},t;\vec{r}\,',t')\left(\partial'_t\rho + \nabla' \cdot \vec{\mathbf{j}}\right) = 0 \end{aligned}$$

via charge conservation: $\partial_t\rho + \nabla \cdot \vec{\mathbf{j}} = 0$.

3.5 Dipole Radiation

We have seen from electrostatics and magnetostatics that

stationary charges	$\Longrightarrow$	electric field
steady current	$\Longrightarrow$	magnetic field
(charges moving in a constant speed)		

We have also learned that Maxwell's equations admit solutions by electromagnetic waves. We now explain

$$\boxed{\text{accelerating charges}} \qquad \Longrightarrow \qquad \boxed{\text{radiation}}$$

In other words, accelerating charges will generate propagating electromagnetic waves.

3.5.1 Spherical Wave

Let us first describe waves which are spreading out from some center (say the origin) and are spherically symmetric. In other words, we consider a solution φ of

$$\Box\varphi = 0$$

and such that $\varphi = \varphi(r,t)$ where $r = |\vec{r}|$. We have seen that for spherically symmetric functions, the wave equation becomes

$$\frac{1}{r}\left(\frac{1}{c^2}\frac{\partial^2}{\partial t^2} - \frac{\partial^2}{\partial r^2}\right)(r\varphi) = 0.$$

A general retarded solution is given by

$$\varphi(r,t) = \frac{f(t - r/c)}{4\pi r}.$$

It represents a spherical wave travelling outward from the origin. Strictly speaking, φ solves the wave equation outside the origin but would have singularity at the origin due to the presence of source. In fact, from the general solution of inhomogeneous wave equation with source S

$$\varphi(r,t) = \frac{1}{4\pi}\int d^3r' \, \frac{S(\vec{r}', t - |\vec{r} - \vec{r}'|/c)}{|\vec{r} - \vec{r}'|},$$

we see that the source is located at the origin given by

$$S(\vec{r}, t) = \delta(\vec{r}) f(t)$$

so in fact φ solves

$$\Box\varphi = \delta(\vec{r}) f(t).$$

Another thing worth mentioning is that the amplitude of φ decays in proportion to $1/r$ as the wave propagates. This is consistent with energy conservation. As the wave propagates, the total energy flux over the sphere of radius r must be the same. The area of the sphere is $4\pi r^2$, and the energy density depends on the square of the wave amplitude, so the amplitude of the wave must decrease as $1/r$.

3.5.2 Electric Dipole Radiation

Consider the following retarded solution

$$\varphi(\vec{r},t) = \frac{1}{4\pi}\int d^3r' \, \frac{f(\vec{r}', t - |\vec{r} - \vec{r}'|/c)}{|\vec{r} - \vec{r}'|}.$$

Let us assume the source f is localized in some region V, and we look φ at a distance far away from the region. Then we have the multipole expansion

$$\frac{1}{|\vec{r} - \vec{r}'|} = \frac{1}{r} + \frac{\vec{r}\cdot\vec{r}'}{r^3} + \cdots \qquad \text{for } r >> 0.$$

We also assume that the source does not vary too fast: the motion of charges and currents are non-relativistic, and they do not change very much over the time that it takes light to cross the region V. This in particular says that the operation $\frac{r'}{c}\partial_t$

would produce something very small. Then we can use $|\vec{r}-\vec{r}'| = r - \frac{\vec{r}\cdot\vec{r}'}{r} + \cdots$ to expand

$$f(\vec{r}', t - |\vec{r}-\vec{r}'|/c) = f\left(\vec{r}', t - r/c + \frac{\vec{r}\cdot\vec{r}'}{rc} + \cdots\right)$$
$$= f(\vec{r}', t - r/c) + \frac{\vec{r}\cdot\vec{r}'}{rc}\partial_t f(\vec{r}', t - r/c) + \cdots .$$

Therefore φ is approximated by

$$\varphi(\vec{r}, t) \simeq \frac{1}{4\pi r}\int d^3r' \left(1 + \frac{\vec{r}\cdot\vec{r}'}{r^2} + \cdots\right)\left(f(\vec{r}', t - r/c) + \frac{\vec{r}\cdot\vec{r}'}{rc}\partial_t f(\vec{r}', t - r/c) + \cdots\right)$$
$$= \frac{1}{4\pi r}\int d^3r' \left[f(\vec{r}', t - r/c) + \frac{\vec{r}\cdot\vec{r}'}{rc}\partial_t f(\vec{r}', t - r/c) + \frac{\vec{r}\cdot\vec{r}'}{r^2} f(\vec{r}', t - r/c) + \cdots\right].$$

Let us apply this to Maxwell's equations with charge distribution ρ and current distribution $\vec{\mathbf{j}}$. The vector potential

$$\vec{\mathbf{A}}(\vec{r}, t) = \frac{\mu_0}{4\pi}\int d^3r' \frac{\vec{\mathbf{j}}(\vec{r}', t - |\vec{r}-\vec{r}'|/c)}{|\vec{r}-\vec{r}'|}$$

has the leading approximation (under the same assumption as above)

$$\vec{\mathbf{A}}(\vec{r}, t) \simeq \frac{\mu_0}{4\pi r}\int d^3r' \,\vec{\mathbf{j}}(\vec{r}', t - r/c).$$

Consider the x-component for example, we have

$$\mathbf{j}_x(\vec{r}', t') = \nabla'\cdot\left(x'\vec{\mathbf{j}}(\vec{r}', t')\right) - x'\nabla'\cdot\vec{\mathbf{j}}(\vec{r}', t') = \nabla'\cdot\left(x'\vec{\mathbf{j}}(\vec{r}', t')\right) + x'\partial_{t'}\rho(\vec{r}', t').$$

Therefore

$$\int d^3r' \,\mathbf{j}_x(\vec{r}', t - r/c) = \int d^3r' \, x'\partial_t\rho(\vec{r}', t - r/c).$$

The y and z components are similar. Thus

$$\frac{\mu_0}{4\pi r}\int d^3r' \,\vec{\mathbf{j}}'(\vec{r}', t - r/c) = \frac{\mu_0}{4\pi r}\frac{d}{dt}\int d^3r' \,\rho(\vec{r}', t - r/c)\vec{r}' = \frac{\mu_0}{4\pi r}\dot{\vec{p}}(t - r/c)$$

where $\vec{p}(t) = \int d^3r' \,\rho(\vec{r}', t)\vec{r}'$ is the electric dipole moment. So the leading approximation of the vector potential is

$$\vec{\mathbf{A}}(\vec{r}, t) \simeq \frac{\mu_0}{4\pi r}\dot{\vec{p}}(t - r/c).$$

This is called the **electric dipole approximation**.

We can compute the approximated magnetic field by

$$\vec{\mathbf{B}} = \nabla\times\vec{\mathbf{A}} \simeq -\frac{\mu_0}{4\pi r^2}\hat{r}\times\dot{\vec{p}}(t - r/c) - \frac{\mu_0}{4\pi rc}\hat{r}\times\ddot{\vec{p}}(t - r/c)$$

where $\hat{r} = \frac{\vec{r}}{r}$. The second term is the leading contribution far away, so

$$\vec{\mathbf{B}} \simeq -\frac{\mu_0}{4\pi r c}\hat{r} \times \ddot{\vec{p}}(t - r/c).$$

We next compute the approximated electric field. The scalar potential

$$\begin{aligned}
\phi(\vec{r}, t) &= \frac{1}{4\pi\varepsilon_0}\int d^3r' \, \frac{\rho(\vec{r}\,', t - |\vec{r} - \vec{r}\,'|/c)}{|\vec{r} - \vec{r}\,'|} \\
&= \frac{1}{4\pi\varepsilon_0 r}\int d^3r' \Big[\rho(\vec{r}\,', t - r/c) + \frac{\vec{r}\cdot\vec{r}\,'}{rc}\partial_t\rho(\vec{r}\,', t - r/c) + \frac{\vec{r}\cdot\vec{r}\,'}{r^2}\rho(\vec{r}\,', t - r/c) + \cdots\Big] \\
&= \frac{1}{4\pi\varepsilon_0 r}\left[Q + \frac{\hat{r}}{c}\cdot\dot{\vec{p}}(t - r/c) + \frac{\hat{r}\cdot\vec{p}(t - r/c)}{r} + \cdots\right].
\end{aligned}$$

Here Q is the total charge. The leading approximation of the electric field is therefore (keeping only the $1/r$-order term)

$$\begin{aligned}
\vec{\mathbf{E}} &= -\nabla\phi - \partial_t\vec{\mathbf{A}} \\
&\simeq \frac{\hat{r}}{4\pi\varepsilon_0 r c}\cdot\ddot{\vec{p}}(t - r/c)\nabla\left(\frac{r}{c}\right) - \frac{\mu_0}{4\pi r}\ddot{\vec{p}}(t - r/c) \\
&= \frac{\mu_0}{4\pi r}\left[\left(\hat{r}\cdot\ddot{\vec{p}}\right)\hat{r} - \ddot{\vec{p}}\,(\hat{r}\cdot\hat{r})\right] \\
&= \frac{\mu_0}{4\pi r}\hat{r}\times\left(\hat{r}\times\ddot{\vec{p}}(t - r/c)\right).
\end{aligned}$$

As a consistency check, we can also use Maxwell's equations. The approximated electric field $\vec{\mathbf{E}}$ can be computed via that of $\vec{\mathbf{B}}$ by

$$\frac{\partial}{\partial t}\vec{\mathbf{E}} = c^2\nabla\times\vec{\mathbf{B}} \simeq \frac{\mu_0}{4\pi r}\hat{r}\times\left(\hat{r}\times\dddot{\vec{p}}\,(t - r/c)\right),$$

so

$$\vec{\mathbf{E}} \simeq \frac{\mu_0}{4\pi r}\hat{r}\times\left(\hat{r}\times\ddot{\vec{p}}(t - r/c)\right).$$

Let us denote this leading electric dipole approximation by

$$\begin{cases}
\vec{\mathbf{B}}_{\text{ED}} = -\dfrac{\mu_0}{4\pi r c}\hat{r}\times\ddot{\vec{p}}(t - r/c) \\
\vec{\mathbf{E}}_{\text{ED}} = \dfrac{\mu_0}{4\pi r}\hat{r}\times\left(\hat{r}\times\ddot{\vec{p}}(t - r/c)\right)
\end{cases}$$

They are both spherical waves and related by

$$\vec{\mathbf{E}}_{\text{ED}} = -c\hat{r}\times\vec{\mathbf{B}}_{\text{ED}}.$$

In particular, both $\vec{\mathbf{E}}_{\text{ED}}$ and $\vec{\mathbf{B}}_{\text{ED}}$ are perpendicular to the propagating direction and decays as $1/r$.

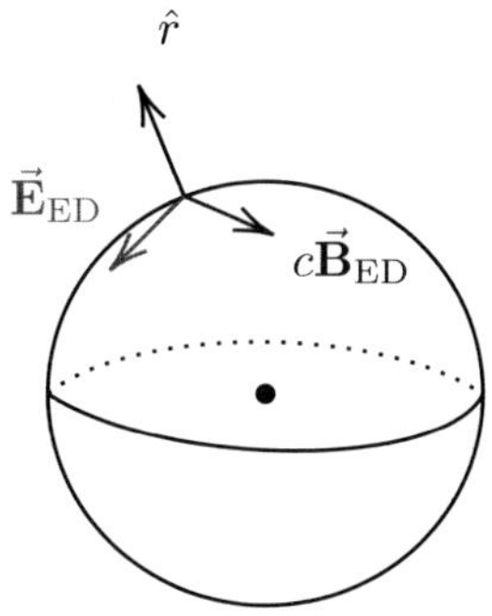

The power radiated from the source can be computed by the Poynting vector

$$\vec{S} = \frac{1}{\mu_0}\vec{\mathbf{E}} \times \vec{\mathbf{B}} \simeq \frac{1}{\mu_0}\vec{\mathbf{E}}_{\mathrm{ED}} \times \vec{\mathbf{B}}_{\mathrm{ED}} = \frac{\mu_0}{16\pi^2 r^2 c}|\hat{r} \times \ddot{\vec{p}}|^2 \hat{r}$$

which points in the same direction as $\vec{r}$. We conclude that oscillating dipole is emitting spherical electromagnetic waves that transporting power radially.

We can also study higher order approximations. For example. the next order are magnetic dipole and electric quadrupole radiations. We will not discuss them in this note.

Let us assume the dipole is oscillating along the z-direction:

$$\vec{p} = p(t)\hat{z}.$$

In spherical coordinate (r, θ, ϕ),

$$|\hat{r} \times \hat{z}| = \sin\theta.$$

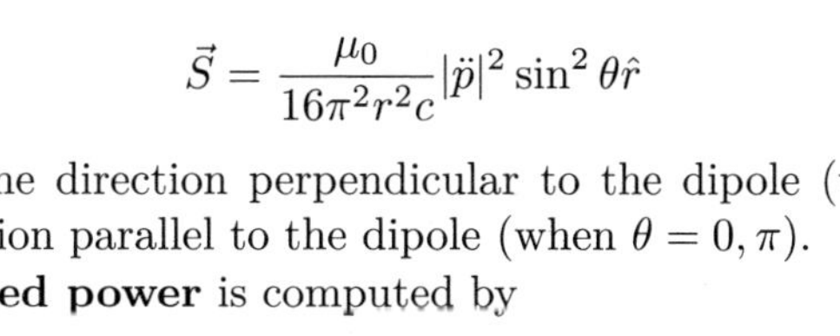

Then

$$\vec{S} = \frac{\mu_0}{16\pi^2 r^2 c}|\ddot{p}|^2 \sin^2\theta \hat{r}$$

which is largest in the direction perpendicular to the dipole (when $\theta = \pi/2$) and smallest in the direction parallel to the dipole (when $\theta = 0, \pi$).

The **total radiated power** is computed by

$$P = \int_{S^2} d\vec{\sigma} \cdot \vec{S} = \frac{\mu_0}{16\pi^2 c}|\ddot{p}|^2 \int_0^{2\pi} d\phi \int_0^{\pi} d\theta \sin^3\theta = \frac{\mu_0}{6\pi c}|\ddot{p}|^2.$$

Example 3.5.1. Consider a particle of charge Q oscillating in the z-direction with frequency ω and amplitude d. The electric dipole moment is

$$\vec{p} = p\cos(\omega t)\hat{z} \qquad \text{where} \qquad p = Qd.$$

Then

$$\ddot{\vec{p}} = -\omega^2 p\cos(\omega t)\hat{z}$$

and the total radiated power is

$$P(t) = \frac{\mu_0 p^2\omega^4}{6\pi c}\cos^2(\omega t).$$

The time-averaged power is

$$\langle P\rangle = \frac{1}{2\pi/\omega}\int_0^{2\pi/\omega} P(t)dt = \frac{\mu_0 p^2\omega^4}{12\pi c}.$$

This is called the **Larmor's formula**.

3.6 Moving Point Charge

Let us consider a point particle with charge q moving in the trajectory

$$\vec{\xi}(t) = \text{ position of } q \text{ at time } t.$$

Since $\vec{\xi}(t)$ is general, the point charge will accelerate along the way and radiate electromagnetic waves. We would like to calculate the electromagnetic fields produced by such a moving point charge.

The charge and current densities are

$$\begin{cases} \rho(\vec{r},t) = q\,\delta^3\left(\vec{r}-\vec{\xi}(t)\right) \\ \vec{\mathbf{j}}\,(\vec{r},t) = q\vec{v}(t)\,\delta^3\left(\vec{r}-\vec{\xi}(t)\right) \end{cases}$$

where $\vec{v}(t) = \dot{\vec{\xi}}(t)$ is the velocity of the particle.

The scalar and vector potentials are then given by

$$\begin{cases} \phi(\vec{r},t) = \dfrac{q}{4\pi\varepsilon_0}\displaystyle\int d^3r'\,\frac{\delta^3\left(\vec{r}'-\vec{\xi}(t-|\vec{r}-\vec{r}'|/c)\right)}{|\vec{r}-\vec{r}'|} \\ \vec{\mathbf{A}}(\vec{r},t) = \dfrac{q\mu_0}{4\pi}\displaystyle\int d^3r'\,\frac{\vec{v}(t-|\vec{r}-\vec{r}'|/c)\,\delta^3\left(\vec{r}-\vec{\xi}(t-|\vec{r}-\vec{r}'|/c)\right)}{|\vec{r}-\vec{r}'|} \end{cases}$$

Let us first consider the scalar potential $\phi(\vec{r},t)$. To deal with the δ-function, let us first rewrite

$$\begin{aligned}\phi(\vec{r},t) &= \frac{q}{4\pi\varepsilon_0}\int dt' \int d^3r' \, \frac{\delta^3\left(\vec{r}\,' - \vec{\xi}(t')\right)\delta\left(t' - (t - |\vec{r} - \vec{r}\,'|/c)\right)}{|\vec{r} - \vec{r}\,'|} \\ &= \frac{q}{4\pi\varepsilon_0}\int dt' \, \frac{\delta\left(t' - t + |\vec{r} - \vec{\xi}(t')|/c\right)}{|\vec{r} - \vec{\xi}(t')|}.\end{aligned}$$

Recall we have the following δ-function relation

$$\delta(f(x)) = \sum_{x_i} \frac{\delta(x - x_i)}{|f'(x_i)|}$$

where the sum is over all roots $f(x_i) = 0$. In fact, this is about change of variable formula:

$$\int g(x)\delta(f(x))df = \sum_{x_i} g(x_i)$$

and

$$\int g(x)\delta(f(x))df = \int g(x)|f'(x)|\delta(f(x))dx.$$

Comparing the above two expressions, we get the above δ-function relation.

Let us now apply this to $\delta(f(t'))$ where

$$f(t') = t' - t + |\vec{r} - \vec{\xi}(t')|/c.$$

A root of $f(t')$ describes a time t' such that

$$|\vec{r} - \vec{\xi}(t')| = c(t - t').$$

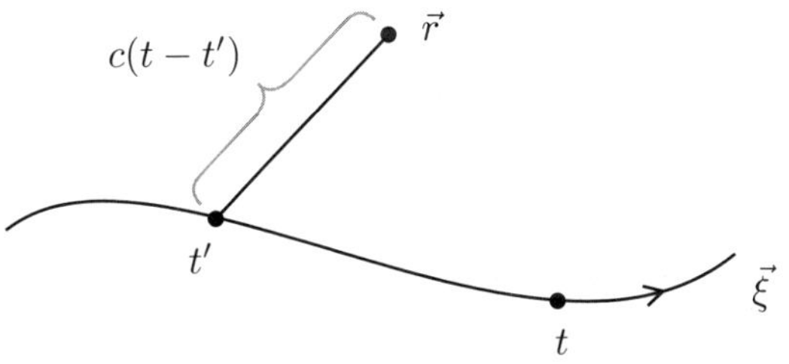

Such a root is called a **retarded time**, where the point $\vec{\xi}(t')$ is at some time t' earlier than t such that the light sent at this point will travel to $\vec{r}$ at time t. Such retarded time is unique. In fact, assume we have two retarded times t_1' and t_2 both solving $f(t_1') = f(t_2') = 0$.

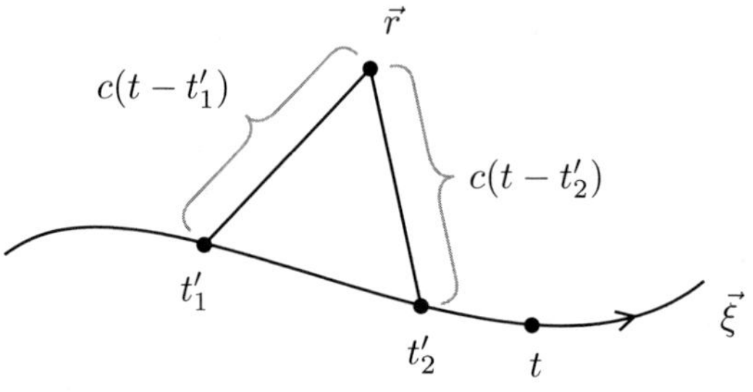

Assume $t'_1 < t'_2 \le t$. Then the length of the curve

$$\text{length}\left(\vec{\xi}\Big|_{[t'_1,t'_2]}\right) \ge c(t-t'_1) - c(t-t'_2) = c(t'_2 - t'_1).$$

We conclude that the speed of the particle $\ge c$. Since no charged particle can travel at the speed of light, this is a contradiction. The existence of retarded time can be achieved by chasing back along the trajectory for long enough until $|\vec{r}-\vec{\xi}(t')| = c(t-t')$ is satisfied since the charged particle travels relatively smaller than the speed of light.

Let us denote t_{ret} for the the retarded time solving the equation

$$t_{\text{ret}} - t + |\vec{r} - \vec{\xi}(t_{\text{ret}})|/c = 0$$

with the understanding that the dependence of t_{ret} on $(\vec{r}, t)$ is implicit when it is clear from the context. To simplify notation, let us also write

$$\vec{R}(t) = \vec{r} - \vec{\xi}(t), \qquad R(t) = |\vec{R}(t)|, \qquad \hat{n}_R = \vec{R}/R, \qquad \vec{\beta}(t) = \vec{v}(t)/c, \qquad \beta(t) = |\vec{\beta}(t)|.$$

$\vec{R}(t)$ represents the relative vector from the point $\vec{\xi}(t)$ of the particle to the point $\vec{r}$ of the field. Then

$$\delta(f(t')) = \frac{1}{1 - \vec{R}(t_{\text{ret}}) \cdot \vec{\beta}(t_{\text{ret}})/R(t_{\text{ret}})}\delta(t' - t_{\text{ret}}).$$

Plugging this into the potential, we find

$$\phi(\vec{r}, t) = \frac{1}{4\pi\varepsilon_0}\left[\frac{q}{R - \vec{R}\cdot\vec{\beta}}\right]_{\text{ret}}.$$

Here $[-]_{\text{ret}}$ means the time variable is evaluated at t_{ret}. By a similar argument, we find the vector potential

$$\mathbf{\vec{A}}(\vec{r}, t) = \frac{\mu_0}{4\pi}\left[\frac{q\vec{v}}{R - \vec{R}\cdot\vec{\beta}}\right]_{\text{ret}}.$$

The above two expressions are called the "**Liénard-Wiechert potentials**" for a moving point charge.

Example 3.6.1 (Point charge of constant velocity). Consider

$$\vec{\xi}(t) = \vec{v}\,t \qquad \text{with} \qquad \vec{v} \text{ constant.}$$

The retarded time is computed via

$$|\vec{r} - \vec{v}\,t_{\text{ret}}| = c(t - t_{\text{ret}}).$$

Geometrically, let C represent the point at time t, and B represent the point at time t_{ret}.

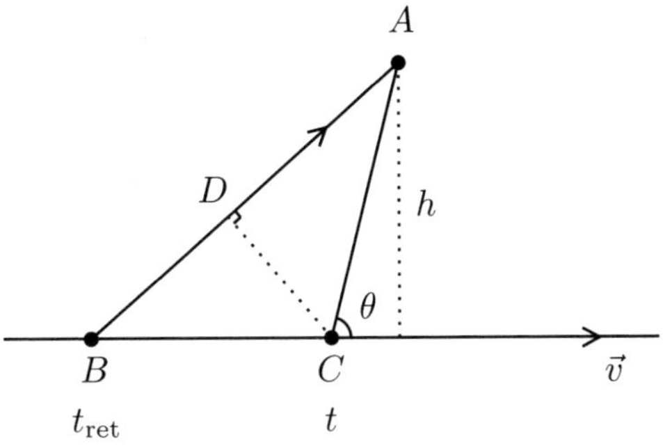

Then

$$|AB| = |\vec{r} - \vec{v}\,t_{\text{ret}}|, \qquad |BC| = v(t - t_{\text{ret}}), \qquad \text{where} \quad v = |\vec{v}\,|.$$

The above equation becomes

$$\frac{|AB|}{|BC|} = \frac{c}{v} = \frac{1}{\beta}.$$

It is not hard to see that

$$\begin{aligned} R(t_{\text{ret}}) - \vec{R}(t_{\text{ret}}) \cdot \vec{\beta} &= |AB| - |BD| = |AD| = \sqrt{|AC|^2 - |CD|^2} \\ &= \sqrt{|AC|^2 - \left(h \cdot \frac{|BC|}{|AB|}\right)^2} \\ &= |AC|\sqrt{1 - \beta^2 \sin^2\theta}. \end{aligned}$$

Therefore

$$\begin{cases} \phi(\vec{r}, t) = \dfrac{1}{4\pi\varepsilon_0} \dfrac{q}{R\sqrt{1 - \beta^2 \sin^2\theta}} \\ \vec{\mathbf{A}}(\vec{r}, t) = \dfrac{\mu_0}{4\pi} \dfrac{\vec{v}}{R\sqrt{1 - \beta^2 \sin^2\theta}} \end{cases}$$

Here $\vec{R} = \vec{r} - \vec{v}\,t$ is the relative vector from the present position of the particle to the field point, and θ is the angle between $\vec{R}$ and $\vec{v}$. Note that for nonrelativistic velocities ($v << c$),

$$\phi(\vec{r}, t) \simeq \frac{1}{4\pi\varepsilon_0} \frac{q}{R}.$$

We now move on to compute the electromagnetic fields

$$\begin{cases} \vec{\mathbf{E}} = -\nabla\phi - \partial_t \vec{\mathbf{A}} \\ \vec{\mathbf{B}} = \nabla \times \vec{\mathbf{A}} \end{cases}$$

Since the retarded time t_{ret} depends on $\vec{r}$ and t implicitly, we work directly with the expression

$$\begin{cases} \phi(\vec{r}, t) = \dfrac{q}{4\pi\varepsilon_0} \displaystyle\int dt' \, \dfrac{\delta\left(t' - t + R(t')/c\right)}{R(t')} \\ \vec{\mathbf{A}}(\vec{r}, t) = \dfrac{q\mu_0}{4\pi} \displaystyle\int dt' \, \dfrac{\vec{v}(t')\delta\left(t' - t + R(t')/c\right)}{R(t')} \end{cases}$$

Using

$$\nabla R = \vec{R}/R = \hat{n}_R, \qquad \nabla\delta\left(t' - t + R(t')/c\right) = \nabla(R/c)\left(-\frac{\partial}{\partial t}\delta\left(t' - t + R(t')/c\right)\right),$$

we find

$$\begin{aligned} \vec{\mathbf{E}} &= -\nabla\phi - \partial_t\vec{\mathbf{A}} \\ &= \frac{q}{4\pi\varepsilon_0}\int dt' \left[\frac{\nabla R(t')}{R(t')^2}\delta\left(t' - t + R(t')/c\right) + \frac{\nabla R(t')}{cR(t')}\frac{\partial}{\partial t}\delta\left(t' - t + R(t')/c\right)\right] \\ &\quad - \frac{q}{4\pi\varepsilon_0}\frac{\partial}{\partial t}\int dt' \, \frac{\vec{v}(t')\delta\left(t' - t + R(t')/c\right)}{c^2 R(t')} \\ &= \frac{q}{4\pi\varepsilon_0}\left[\frac{\vec{R}}{R^3 g}\right]_{\text{ret}} + \frac{q}{4\pi\varepsilon_0 c}\frac{\partial}{\partial t}\left[\frac{\vec{R}/R - \vec{v}/c}{Rg}\right]_{\text{ret}} \\ &= \frac{q}{4\pi\varepsilon_0}\left[\frac{\hat{n}_R}{R^2 g}\right]_{\text{ret}} + \frac{q}{4\pi\varepsilon_0}\frac{\partial}{\partial t}\left[\frac{\hat{n}_R - \vec{\beta}}{Rgc}\right]_{\text{ret}}. \end{aligned}$$

where $g = 1 - \hat{n}_R \cdot \vec{\beta} = 1 - \frac{\vec{R}\cdot\vec{v}}{Rc}$. To compute t-derivative term, we use

$$t_{\text{ret}} - t + R(t_{\text{ret}})/c = 0$$

to find

$$\frac{dt}{dt_{\text{ret}}} = 1 + \frac{1}{c}\frac{\partial R}{\partial t}(t_{\text{ret}}) = g(t_{\text{ret}}).$$

From $\vec{R} = \vec{r} - \vec{\xi}(t)$, we have

$$\frac{\partial\vec{R}}{\partial t} = -\vec{v} = -c\vec{\beta}, \qquad \frac{\partial R}{\partial t} = -c\hat{n}_R \cdot \vec{\beta},$$

$$\begin{aligned} \frac{\partial\hat{n}_R}{\partial t} &= \frac{1}{R}\frac{\partial\vec{R}}{\partial t} - \frac{\vec{R}}{R^3}\left(\vec{R}\cdot\frac{\partial\vec{R}}{\partial t}\right) = \frac{1}{R}\left(\frac{\partial\vec{R}}{\partial t} - \left(\hat{n}_R\cdot\frac{\partial\vec{R}}{\partial t}\right)\hat{n}_R\right) = \frac{c}{R}\left(\left(\hat{n}_R\cdot\vec{\beta}\right)\hat{n}_R - \vec{\beta}\right) \\ &= \frac{c}{R}\hat{n}_R \times \left(\hat{n}_R \times \vec{\beta}\right) = \frac{c}{R}\left(\hat{n}_R - \vec{\beta}\right) - \frac{c}{R}g\hat{n}_R, \end{aligned}$$

$$\frac{\partial Rg}{\partial t} = \frac{\partial}{\partial t}\left(R - \vec{R}\cdot\vec{\beta}\right) = -c\hat{n}_R\cdot\vec{\beta} + c\beta^2 - \vec{R}\cdot\frac{d\vec{\beta}}{dt} = -c\hat{n}_R\cdot\vec{\beta} + c\beta^2 - R\,\hat{n}_R\cdot\frac{d\vec{\beta}}{dt}.$$

This enables us to compute

$$\begin{aligned}\frac{\partial}{\partial t}\left[\frac{\hat{n}_R-\vec{\beta}}{Rgc}\right]_{\text{ret}} &= \frac{dt_{\text{ret}}}{dt}\frac{\partial}{\partial t_{\text{ret}}}\left[\frac{\hat{n}_R-\vec{\beta}}{Rgc}\right]_{\text{ret}} = \frac{1}{g\left(t_{\text{ret}}\right)}\left[\frac{\partial}{\partial t}\left(\frac{\hat{n}_R-\vec{\beta}}{Rgc}\right)\right]_{\text{ret}} \\ &= \left[\frac{1}{g}\frac{\partial}{\partial t}\left(\frac{\hat{n}_R-\vec{\beta}}{Rgc}\right)\right]_{\text{ret}}\end{aligned}$$

and

$$\begin{aligned}\vec{\mathbf{E}} &= \frac{q}{4\pi\varepsilon_0}\left[\frac{\hat{n}_R}{R^2g} + \frac{\frac{\partial}{\partial t}\left(\hat{n}_R-\vec{\beta}\right)}{Rg^2c} - \frac{\hat{n}_R-\vec{\beta}}{R^2g^3c}\frac{\partial}{\partial t}(Rg)\right]_{\text{ret}} \\ &= \frac{q}{4\pi\varepsilon_0}\left[\frac{\hat{n}_R}{R^2g} + \frac{\hat{n}_R-\vec{\beta}}{R^2g^2} - \frac{\hat{n}_R}{R^2g} - \frac{d\vec{\beta}/dt}{Rg^2c} - \frac{\hat{n}_R-\vec{\beta}}{R^2g^3c}\left(-c\hat{n}_R\cdot\vec{\beta}+c\beta^2 - \vec{R}\cdot\frac{d\vec{\beta}}{dt}\right)\right]_{\text{ret}} \\ &= \frac{q}{4\pi\varepsilon_0}\left[\frac{\hat{n}_R-\vec{\beta}}{R^2g^3}\left(g+\hat{n}_R\cdot\vec{\beta}-\beta^2\right) + \frac{\hat{n}_R-\vec{\beta}}{Rg^3c}\left(\hat{n}_R\cdot\frac{d\vec{\beta}}{dt}\right) - \frac{d\vec{\beta}/dt}{Rg^2c}\right]_{\text{ret}} \\ &= \frac{q}{4\pi\varepsilon_0}\left[\frac{\left(\hat{n}_R-\vec{\beta}\right)(1-\beta^2)}{R^2g^3} + \frac{\hat{n}_R\times\left(\left(\hat{n}_R-\vec{\beta}\right)\times\frac{d\vec{\beta}}{dt}\right)}{Rg^3c}\right]_{\text{ret}}.\end{aligned}$$

That is,

$$\vec{\mathbf{E}} = \frac{q}{4\pi\varepsilon_0}\left[\frac{\left(\hat{n}_R-\vec{\beta}\right)(1-\beta^2)}{R^2g^3} + \frac{\hat{n}_R\times\left(\left(\hat{n}_R-\vec{\beta}\right)\times\frac{d\vec{\beta}}{dt}\right)}{Rg^3c}\right]_{\text{ret}}.$$

A similar computation leads to

$$\vec{\mathbf{B}} = \frac{1}{c}\left[\hat{n}_R\right]_{\text{ret}}\times\vec{\mathbf{E}}.$$

These are called **Liénard-Wiechert electric and magnetic fields**.

The above expression decomposes the electric and magnetic field into a velocity field and an acceleration field: $\vec{\mathbf{E}} = \vec{\mathbf{E}}_v + \vec{\mathbf{E}}_a$ where

$$\begin{cases}\vec{\mathbf{E}}_v = \dfrac{q}{4\pi\varepsilon_0}\left[\dfrac{\left(\hat{n}_R-\vec{\beta}\right)(1-\beta^2)}{R^2g^3}\right]_{\text{ret}} \\ \vec{\mathbf{E}}_a = \dfrac{q}{4\pi\varepsilon_0}\left[\dfrac{\hat{n}_R\times\left(\left(\hat{n}_R-\vec{\beta}\right)\times\frac{d\vec{\beta}}{dt}\right)}{Rg^3c}\right]_{\text{ret}}\end{cases}$$

$\vec{\mathbf{E}}_v$ only depends on the velocity $\vec{\beta}$ and decays as $1/R^2$ in space. This field and its energy is attached to the particle. $\vec{\mathbf{E}}_a$ contains $\dot{\vec{\beta}}$ which is about the acceleration. It decays as $1/R$ in space, and therefore contains surface energy that propagates to infinity. This is the radiation field which contain energy that radiates from the particle to the space faraway.

Example 3.6.2. We again look at a point charge of constant velocity with $\vec{\xi}(t) = \vec{v}t$.

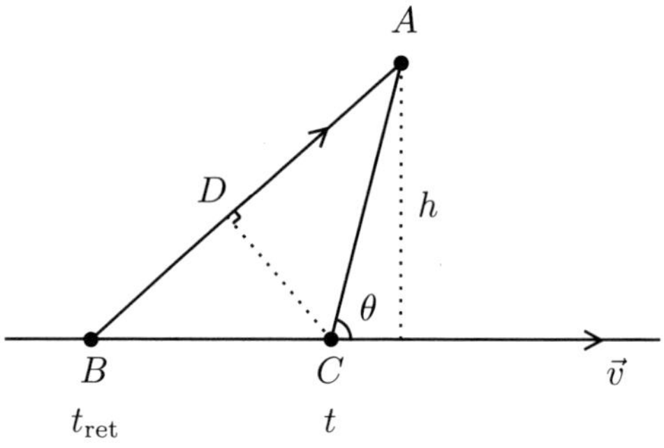

The particle is not accelerating and

$$\vec{\mathbf{E}} = \vec{\mathbf{E}}_v = \frac{q}{4\pi\varepsilon_0}\left[\frac{\left(\hat{n}_R - \vec{\beta}\right)(1-\beta^2)}{R^2 g^3}\right]_{\text{ret}} = \frac{q}{4\pi\varepsilon_0}\left[\frac{\left(\vec{R} - R\vec{\beta}\right)(1-\beta^2)}{R^3 g^3}\right]_{\text{ret}}.$$

We have seen in the previous example that

$$[Rg]_{\text{ret}} = R\left(1-\beta^2\sin^2\theta\right)^{1/2},$$

$$\left[\vec{R} - R\vec{\beta}\right]_{\text{ret}} = \overrightarrow{BA} - \overrightarrow{BC} = \overrightarrow{CA} = \vec{R}.$$

It follows that

$$\vec{\mathbf{E}} = \frac{q}{4\pi\varepsilon_0}\frac{1-v^2/c^2}{\left(1-v^2\sin^2\theta/c^2\right)^{3/2}}\frac{\vec{R}}{R^3}$$

which remains in the radial direction, but is stronger in the direction perpendicular to the moving direction and weaker in the direction parallel to the moving direction.

For example, assume the particle is moving along x-direction with

$$\vec{\xi}(t) = (vt, 0, 0).$$

Then the above formula leads to the explicit expression in components $\vec{\mathbf{E}} = (\mathbf{E}_x, \mathbf{E}_y, \mathbf{E}_z)$

$$\begin{cases} \mathbf{E}_x = \dfrac{q}{4\pi\varepsilon_0}\dfrac{\gamma(x-vt)}{\left(\gamma^2(x-vt)^2+y^2+z^2\right)^{3/2}} \\ \mathbf{E}_y = \dfrac{q}{4\pi\varepsilon_0}\dfrac{\gamma y}{\left(\gamma^2(x-vt)^2+y^2+z^2\right)^{3/2}} \\ \mathbf{E}_z = \dfrac{q}{4\pi\varepsilon_0}\dfrac{\gamma z}{\left(\gamma^2(x-vt)^2+y^2+z^2\right)^{3/2}} \end{cases}$$

where $\gamma = \frac{1}{\sqrt{1-v^2/c^2}}$. This expression has a very suggestive relativistic meaning. Indeed, we will show in Example 5.2.5 that this formula can be simply obtained from the Lorentz boost along x-direction of the fields of a stationary point charge.

3.7 Scattering

We briefly discuss the basic idea behind the scattering of electromagnetic waves. Its physics contains several steps:

Incident electromagnetic wave hits the particle

↓

Particle oscillates and accelerates

↓

Radiate scattering electromagnetic wave

The ratio

$$\sigma = \frac{\text{scattered power}}{\text{incident power per unit area}}$$

is called the **cross-section** for scattering.

Thomson Scattering

Consider an incoming plane wave $\vec{\mathbf{E}} = \vec{\mathbf{E}}_0 \cos(\vec{k}\cdot\vec{r} - \omega t)$ interacting with a free particle of mass m and charge q. The equation of motion is

$$m\ddot{\vec{r}} = q\left(\vec{\mathbf{E}} + \dot{\vec{r}} \times \vec{\mathbf{B}}\right).$$

When the particle speed is non-relativistic, $|\dot{\vec{r}}|/c << 1$,

$$|\dot{\vec{r}} \times \vec{\mathbf{B}}| = \left|\frac{\dot{\vec{r}}}{c} \times c\vec{\mathbf{B}}\right| << |c\vec{\mathbf{B}}| = |\vec{\mathbf{E}}|,$$

so we can neglect the magnetic Lorentz force. In the non-relativistic limit, we can also assume the oscillation of the particle is small comparing with the wave length. Under these simplification, we have

$$m\ddot{\vec{r}} = q\vec{\mathbf{E}}_0 \cos\omega t$$

which can be solved by

$$\vec{r} = -\frac{q\vec{\mathbf{E}}_0}{m\omega^2}\cos\omega t.$$

We use the electric dipole approximation and the Larmor's formula as discussed in Section 3.5.2 to compute the time-averaged radiation power

$$\langle P_{\text{rad}}\rangle = \frac{\mu_0\left[q\left(\frac{q\mathbf{E}_0}{m\omega^2}\right)\right]^2\omega^4}{12\pi c} = \frac{\mu_0 q^4 \mathbf{E}_0^2}{12\pi m^2 c}.$$

On the other hand, we have computed the Poynting vector for the plane wave (Example 3.1.5)

$$\vec{S}_{\text{inc}} = \frac{\mathbf{E}_0^2}{c\mu_0} \cos^2\left(\vec{k}\cdot\vec{r} - \omega t\right)\hat{k}$$

whose time average over a single period $2\pi/\omega$ is

$$\left\langle \widehat{S}_{\text{inc}} \right\rangle = \frac{\mathbf{E}_0^2}{2c\mu_0}.$$

The cross-section is given by

$$\sigma = \frac{\langle P_{\text{rad}} \rangle}{\left\langle \widehat{S}_{\text{inc}} \right\rangle} = \frac{\mu_0^2 q^4}{6\pi m^2}.$$

Note that the cross-section of Thomson scattering does not depend on the frequency ω. All wave lengths of light are scattered equally.

Rayleigh Scattering

Rayleigh scattering describes the scattering of electromagnetic waves on a neutral molecule or a small dielectric object. The object exhibits electric dipole polarization under the effect of the electric field

$$\vec{P} = \alpha\vec{\mathbf{E}}.$$

We assume linear polarization so α is a constant. For the plane wave $\vec{\mathbf{E}} = \vec{\mathbf{E}}_0 \cos(\vec{k}\cdot\vec{r} - \omega t)$ as above, Larmor's formula gives the time-averaged radiation power

$$\langle P_{\text{rad}} \rangle = \frac{\mu_0 \alpha^2 \mathbf{E}_0^2 \omega^4}{12\pi c}.$$

The cross-section is given by

$$\sigma = \frac{\langle P_{\text{rad}} \rangle}{\left\langle \widehat{S}_{\text{inc}} \right\rangle} = \frac{\mu_0^2 \alpha^2 \omega^4}{6\pi}$$

which is stronger for high frequencies or short wave lengths. This explains blue sky in the day time and red sky at sunset since

$$\lambda_{\text{blue}} < \lambda_{\text{red}}.$$

During the day, we look away from the sun and see light that has scattered by the atmosphere. At sunset, we look directly at the sun and see light that remains from the scattering.

Chapter 4

$U(1)$ Gauge Theory

There are four fundamental interactions in nature: the weak nuclear force, the strong nuclear force, the electromagnetic force, and the gravity. One landmark of modern physics is to realise that all these interactions have a common feature in terms of gauge principle.

The notion of gauge transformation and gauge invariance was introduced by Hermann Weyl in his attempt to unify gravitation and electromagnetism via a geometric framework. Weyl emphasized the role of gauge invariance as a symmetric principle, and his proposal forms the foundation of what is now known as gauge theory.

The modern aspect of gauge theory has rich content both in mathematics and physics. We are not intended to present a full overview of this beautiful theory. Instead, we aim to put hands on one basic example, the electromagnetism, as a $U(1)$-gauge theory from the modern geometric perspective. We assume basic knowledge on the notion of manifolds, and readers can consult [10, Chapter 3] in this series for preliminary geometric backgrounds. We will give a self-contained discussion on the geometry of fiber bundles in order to understand the bridge between the physical content in the first half of this note and the mathematical content that constitutes the modern framework. We will briefly mention and comment on generalizations, such as non-abelian gauge theory or Yang-Mills theory, at certain steps along the way.

4.1 Fiber Bundle

4.1.1 Fiber Bundle

The notion of **fiber bundle** describes a family of geometric object (called the **fiber**) varying with respect to a parameter space (called the **base**). Precisely, it consists of the following geometric data:

- a manifold B, called the **base manifold**
- a manifold F, called the **fiber**

- a manifold E, called the **total space**
- a Lie group G, called the **structure group**.

Their relationships are described by the following:

① a surjective map $\pi : E \to B$ which is locally trivial with fiber F. This means that there exists a covering $\{\mathfrak{U}_\alpha\}$ of B such that $\pi^{-1}(\mathfrak{U}_\alpha)$ is diffeomorphic to the product $\mathfrak{U}_\alpha \times F$ via

$$\begin{array}{ccc} \pi^{-1}(\mathfrak{U}_\alpha) & \xrightarrow{\quad \varphi_\alpha \quad} & \mathfrak{U}_\alpha \times F \\ & {\scriptstyle \pi} \searrow \quad \swarrow {\scriptstyle \pi_\alpha} & \\ & \mathfrak{U}_\alpha & \end{array}$$

and this diagram is commutative, i.e. $\pi_\alpha \circ \varphi_\alpha = \pi$. Here $\pi_\alpha : \mathfrak{U}_\alpha \times F \to \mathfrak{U}_\alpha$ is the projection to the $\mathfrak{U}_\alpha$-factor. This property says that for each $x \in \mathfrak{U}_\alpha$, φ_α defines a diffeomorphism

$$\varphi_\alpha : \pi^{-1}(x) \xrightarrow{\sim} F$$

so every preimage $\pi^{-1}(x)$ of a point $x \in B$ is diffeomorphic to F. The total space looks like

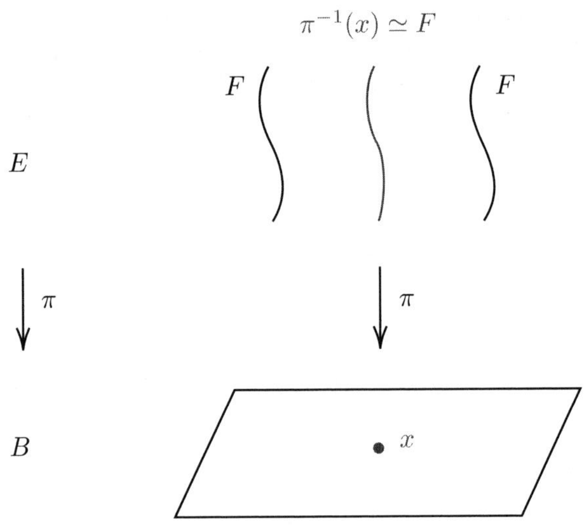

so $\pi : E \to B$ can be viewed as a family of the fiber manifold F varying over the base B. The locally trivial condition says that locally it can be parameterized as a trivial family. However, globally E may not be the same as $B \times F$: nontrivial topological phenomenon can arise.

② G is equipped with an action on F

$$\rho : G \times F \longrightarrow F$$

which we denote simply by $g \cdot u \in F$ for the action of $g \in G$ on $u \in F$. Each g gives a way to identify F with itself. Equivalently, we have a group homomorphism

$$\rho : G \longrightarrow \mathrm{Diff}(F)$$

where $\mathrm{Diff}(F)$ denotes the diffeomorphism group of F.

③ We demand compatibility between ① and ②. Precisely, let

$$\begin{cases} \varphi_\alpha : \pi^{-1}(\mathfrak{U}_\alpha) \longrightarrow \mathfrak{U}_\alpha \times F \\ \varphi_\beta : \pi^{-1}(\mathfrak{U}_\beta) \longrightarrow \mathfrak{U}_\beta \times F \end{cases}$$

be two local trivializations as in ①. Then on the intersection $\mathfrak{U}_{\alpha\beta} := \mathfrak{U}_\alpha \cap \mathfrak{U}_\beta$, we have

$$\begin{array}{ccccc} \mathfrak{U}_{\alpha\beta} \times F & \xleftarrow{\varphi_\alpha} & \pi^{-1}(\mathfrak{U}_{\alpha\beta}) & \xrightarrow{\varphi_\beta} & \mathfrak{U}_{\alpha\beta} \times F \\ & \searrow & \downarrow & \swarrow & \\ & & \mathfrak{U}_{\alpha\beta} & & \end{array}$$

By definition, the composition

$$\varphi_{\beta\alpha} := \varphi_\beta \circ {\varphi_\alpha}^{-1} : \mathfrak{U}_{\alpha\beta} \times F \longrightarrow \mathfrak{U}_{\alpha\beta} \times F$$

is a fiberwise diffeomorphism, i.e. for each $x \in \mathfrak{U}_{\alpha\beta}$, $\varphi_{\beta\alpha}$ maps $\{x\} \times F$ to $\{x\} \times F$ and defines a diffeomorphism of F under the canonical identification $\{x\} \times F \simeq F$. Equivalently, we can write

$$\varphi_{\beta\alpha} : \mathfrak{U}_\alpha \cap \mathfrak{U}_\beta \longrightarrow \mathrm{Diff}(F).$$

Such $\varphi_{\beta\alpha}$ is called the **translation function**, which characterizes the difference of identifying the fiber $\pi^{-1}(x) \simeq F$ under two trivializations. In fact, for $x \in \mathfrak{U}_{\alpha\beta}$, we have

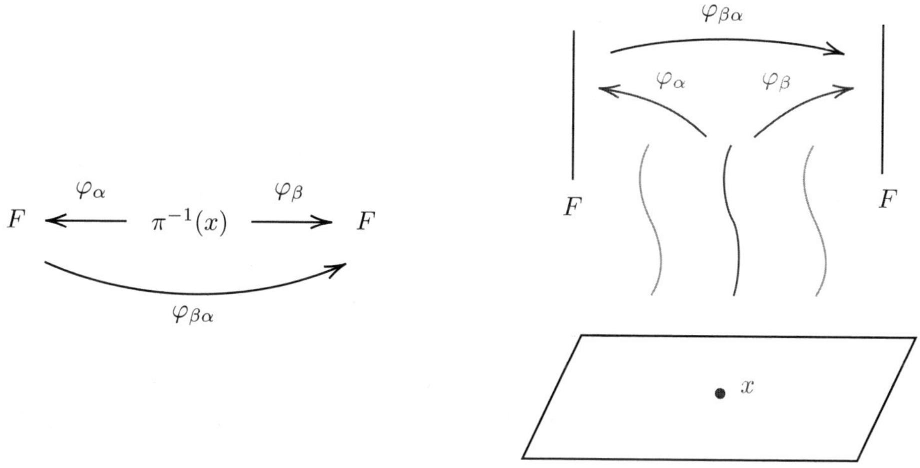

Then we require that each $\varphi_{\beta\alpha}(x)$ is realized by a group action by $g \in G$ on F:

$$\varphi_{\beta\alpha}(x) \cdot u = g(x) \cdot u \qquad \forall u \in F$$

for some $g(x) \in G$. We say the transformation has the structure captured by the Lie group G, and this is why G is called the **structure group**. We will simply write $\varphi_{\beta\alpha}(x) \in G$. Then the transformation function is required to be expressed as

$$\varphi_{\beta\alpha} : \mathfrak{U}_\alpha \cap \mathfrak{U}_\beta \longrightarrow G$$

with the understanding that its actual transformation on the fiber F is realized via the G-action

$$\mathfrak{U}_\alpha \cap \mathfrak{U}_\beta \longrightarrow G \xrightarrow{\rho} \mathrm{Diff}(F).$$

This requirement clearly puts further constraint on the trivializations $\{\mathfrak{U}_\alpha, \varphi_\alpha\}$, and in this case we say the fiber bundle has structure group G.

We usually denote a fiber bundle by

$$\begin{array}{ccc} F & \longrightarrow & E \\ & & \downarrow{\scriptstyle \pi} \\ & & B \end{array}$$

and specify the structure group G in the context when we need to.

Definition 4.1.1. $\pi : E \to B$ is called a **trivial bundle** if there is a global trivialization

$$\begin{array}{ccc} E & \xrightarrow{\quad \simeq \quad} & B \times F \\ & \searrow \quad \swarrow & \\ & B & \end{array}$$

so we can identify E as the product $B \times F$.

Theorem 4.1.2. *If B is contractible, then any fiber bundle E over B is a trivial bundle.*

In particular, any fiber bundle over $\mathbb{R}^n$ is trivial.

Definition 4.1.3. A (smooth) **section** of the fiber bundle $\pi : E \to B$ over a subspace $\mathfrak{U} \xhookrightarrow{j} B$ is a (smooth) map $s : \mathfrak{U} \to E$ such that the following diagram is commutative

$$\begin{array}{ccc} & & E \\ & \overset{s}{\nearrow} & \downarrow{\scriptstyle \pi} \\ \mathfrak{U} & \xhookrightarrow{\ j\ } & B \end{array} \qquad \text{i.e.} \quad j = \pi \circ s.$$

In other words, a section s over $\mathfrak{U}$ assigns every point $x \in \mathfrak{U}$ an element $s(x)$ of the fiber $\pi^{-1}(x)$. We will mainly consider smooth sections without further specification in this note.

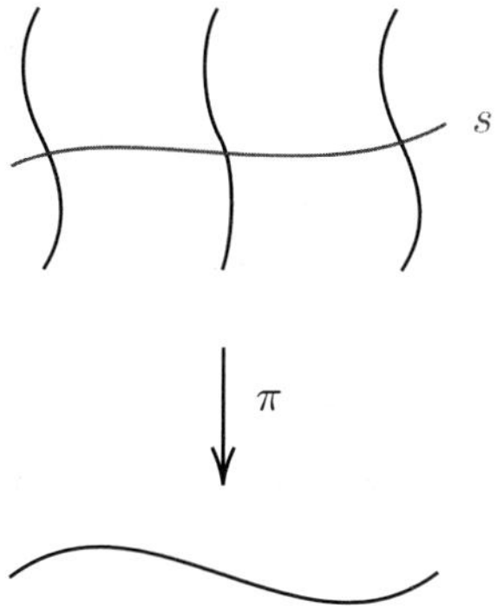

We denote

$$\Gamma(\mathcal{U}, E) = \{\text{sections over } \mathcal{U}\}.$$

Elements of $\Gamma(B, E)$ are also called **global sections**.

Example 4.1.4. If $E = B \times F$ is trivial, then

$$\Gamma(B, E) = \mathrm{Map}(B, F)$$

can be identified with smooth maps from B to F.

$$\begin{aligned} s : B &\longrightarrow E = B \times F \\ b &\longmapsto (b, f(b)) \end{aligned}$$

for $f : B \to F$.

Example 4.1.5 (Möbius strip). Möbius strip is obtained by gluing the two ends of a strip with a half-twist, as shown below

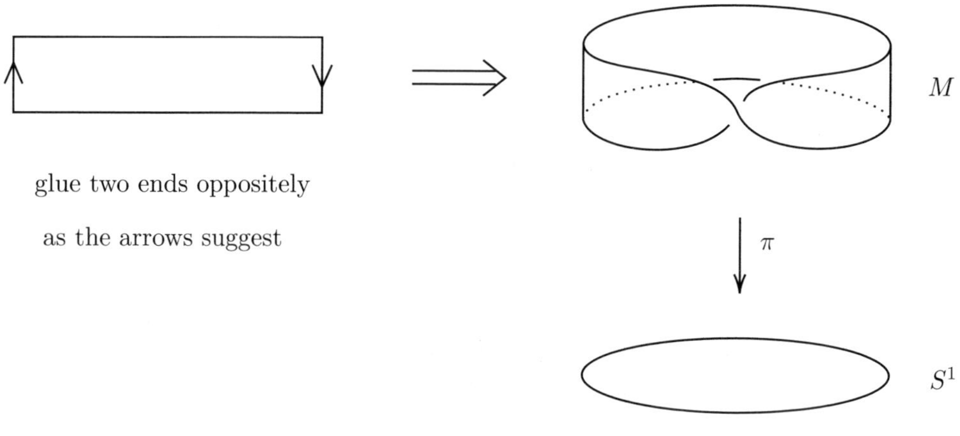

Mathematically, the Möbius strip is described by the quotient

$$M = (I \times I)/\sim = \{(t,x)\,|\,t,x \in I = [0,1]\}/\sim$$

where the equivalence relation is to identify

$$(0,x) \sim (1, 1-x) \qquad \text{for } x \in I = [0,1].$$

It can be viewed as a fiber bundle over S^1

$$\begin{aligned} \pi : M &\longrightarrow S^1 \\ (t,x) &\longmapsto t \end{aligned}$$

where we identify $S^1 = I/\{0,1\}$.

It is not hard to see that for any small open interval $\mathfrak{U} \subset S^1$, $\pi^{-1}(\mathfrak{U})$ can be trivialized as $\mathfrak{U} \times I$ by "straightening". We leave the details to the reader. So

$$\begin{array}{ccc} I & \longrightarrow & M \\ & & \downarrow \pi \\ & & S^1 \end{array}$$

is a fiber bundle with fiber $I = [0,1]$.

By construction, it is clear that global sections of this fiber bundle can be identified with

$$\Gamma(S^1, M) = \{f : [0,1] \to [0,1]\,|\,f(1) = 1 - f(0)\}.$$

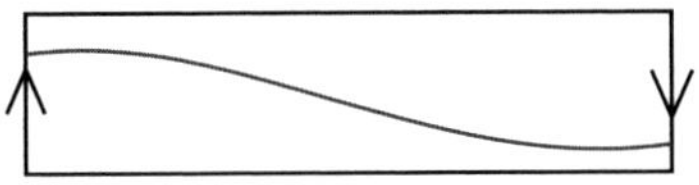

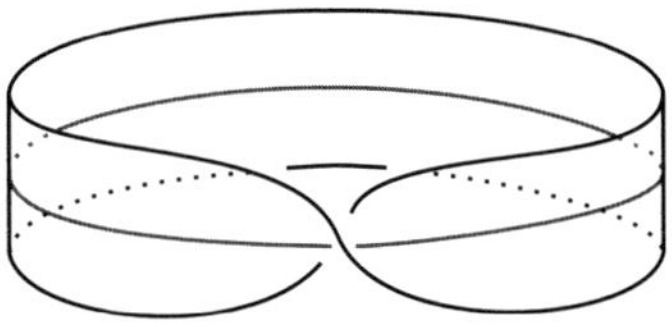

The Möbius strip is a nontrivial fiber bundle. To see this, assume $M \simeq S^1 \times I$ is trivial, then we can find two global sections f_1, f_2 such that their images in M do not intersect. On the other hand, as described above, f_1 and f_2 can be identified with two functions

$$f_i : [0,1] \longrightarrow [0,1]$$

such that $f_i(1) = 1 - f_i(0)$. Assume $f_2(0) > f_1(0)$. Then $f_2(1) < f_1(1)$. By Intermediate Value Theorem, there must be some point t such that

$$f_2(t) = f_1(t).$$

So these two sections must intersect. Contradiction.

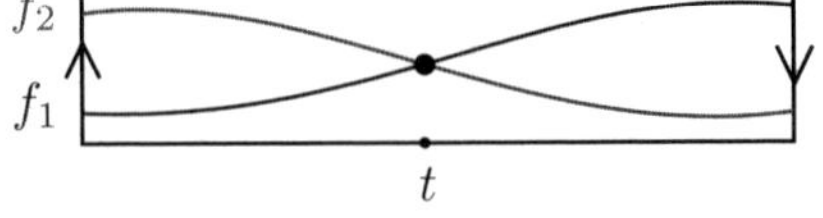

This proves that $\pi : M \to S^1$ is nontrivial.

Example 4.1.6 (Hopf fibration). Consider the map

$$\pi : S^3 \longrightarrow S^2$$

defined as follows. Let us use complex numbers to identify

$$S^3 = \{(z_1, z_2) \in \mathbb{C}^2 \mid |z_1|^2 + |z_2|^2 = 1\}$$
$$S^2 = \mathbb{C}P^1 = \left(\mathbb{C}^2 - \{0\}\right) / \sim$$

A point of $S^2 = \mathbb{C}P^1$ is described by homogeneous coordinates $[z_1, z_2]$, where we identify $[\lambda z_1, \lambda z_2] = [z_1, z_2]$ for all $\lambda \in \mathbb{C}^*$. Then the map π is expressed as

$$\begin{aligned} \pi : S^3 &\longrightarrow S^2 \\ (z_1, z_2) &\longmapsto [z_1, z_2] \end{aligned}$$

For each point $p \in S^2$, we have $\pi^{-1}(p) \simeq S^1$. For example, consider the north pole $[1, 0]$, then

$$\pi^{-1}([1, 0]) = \{(z, 0) \mid z \in \mathbb{C},\ |z|^2 = 1\}.$$

It can be checked that

$$\begin{array}{ccc} S^1 & \longrightarrow & S^3 \\ & & \big\downarrow \pi \\ & & S^2 \end{array}$$

is a fiber bundle with fiber S^1. This is called the "**Hopf fibration**". This is a nontrivial fibration since S^3 and $S^2 \times S^1$ are topologically different. For example,

$$\begin{cases} \pi_1(S^3) = 1 \\ \pi_1(S^2 \times S^1) = \mathbb{Z} \end{cases}$$

i.e. any loop in S^3 can be shrinked continuously to a point, but a loop wrapping along the S^1 factor of $S^2 \times S^1$ can not.

4.1.2 Vector Bundle

Definition 4.1.7. A **real vector bundle**, or simply vector bundle, is a fiber bundle E with fiber $F = \mathbb{R}^m$ and structure group $G \subset \mathrm{GL}_m(\mathbb{R})$. The integer m is called the **rank** of the vector bundle, and we denote

$$\mathrm{rank}(E) = m.$$

Similarly, a **complex vector bundle** of rank m is a fiber bundle with fiber $\mathbb{C}^m$ and structure group $G \subset \mathrm{GL}_m(\mathbb{C})$.

One important fact about vector bundle is that each fiber $\pi^{-1}(x)$ for $x \in B$ is a linear vector space, so we can add elements of $\pi^{-1}(x)$ and multiply by a scalar. In fact, let $\{\mathfrak{U}_\alpha, \varphi_\alpha\}$ be a trivialization of the vector bundle E. Assume $x \in \mathfrak{U}_\alpha$. Then we can use φ_α to identify

$$\varphi_\alpha : \pi^{-1}(x) \longrightarrow \mathbb{R}^m.$$

This allows us to define the linear structure on $\pi^{-1}(x)$ by

$$\lambda_1 s_1 + \lambda_2 s_2 := \varphi_\alpha^{-1}(\lambda_1 \varphi_\alpha(s_1) + \lambda_2 \varphi_\alpha(s_2))$$

where $\lambda_i \in \mathbb{R}$, $s_1 \in \pi^{-1}(x)$. It does not depend on the choice of the local trivialization. Let $x \in \mathfrak{U}_\beta$ and

$$\varphi_\beta : \pi^{-1}(x) \longrightarrow \mathbb{R}^m$$

be another identification of the fiber $\pi^{-1}(x)$ via a different trivialization φ_β. Then the transition

$$\begin{array}{ccc} \mathbb{R}^m & \xrightarrow{\varphi_\alpha^{-1}} & \pi^{-1}(x) \\ & \searrow^{g} & \downarrow \varphi_\beta \\ & & \mathbb{R}^m \end{array}$$

is a linear map $g \in \mathrm{GL}_m(\mathbb{R})$. Therefore

$$g\left(\lambda_1 \varphi_\alpha(s_1) + \lambda_2 \varphi_\alpha(s_2)\right) = \lambda_1 g(\varphi_\alpha(s_1)) + \lambda_2 g(\varphi_\alpha(s_2)),$$

i.e.

$$\varphi_\alpha^{-1}\left(\lambda_1 \varphi_\alpha(s_1) + \lambda_2 \varphi_\alpha(s_2)\right) = \varphi_\beta^{-1}\left(\lambda_1 \varphi_\beta(s_1) + \lambda_2 \varphi_\beta(s_2)\right).$$

This implies the linear operation $\lambda_1 \varphi_\alpha(s_1) + \lambda_2 \varphi_\alpha(s_2)$ is intrinsically defined on the fiber of a vector bundle, and it does not depend on the choice of the local trivialization.

Similarly, let $\mathfrak{U} \subset B$ be an open subset. Then the space of sections of the vector bundle on $\mathfrak{U}$ has a structure of $C^\infty(U)$-module:

$$\Gamma(\mathfrak{U}, E) \text{ is a } C^\infty(U)\text{-module.}$$

For any two sections $s_1, s_2 \in \Gamma(\mathfrak{U}, E)$ and any two functions $f_1, f_2 \in C^\infty(U)$, we have

$$s = f_1 s_1 + f_2 s_2 \in \Gamma(\mathfrak{U}, E)$$

by defining the value of $s(x)$ at each $x \in \mathfrak{U}$ via

$$s(x) = f_1(x) s_1(x) + f_2(x) s_2(x).$$

In particular, global sections $\Gamma(B, E)$ is a $C^\infty(B)$-module.

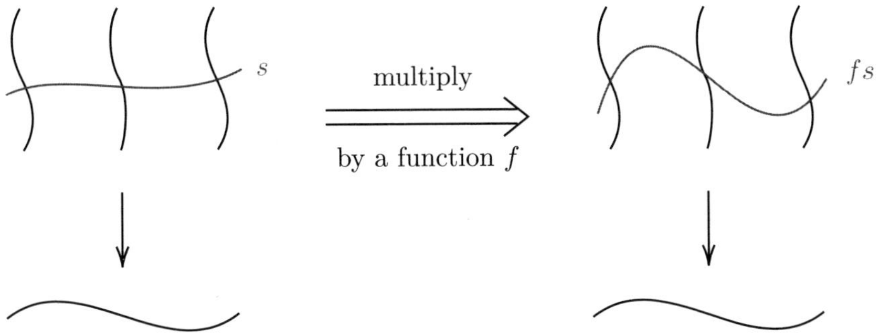

Definition 4.1.8. Let $E \xrightarrow{\pi} B$ be a vector bundle of rank m, and $\mathfrak{U} \subset B$ be an open subset. A set of m sections $\{s_1, \cdots, s_m\}$ of E over $\mathfrak{U}$ is said to be a **frame** of E over $\mathfrak{U}$ if

$$\{s_1(x), \cdots, s_m(x)\} \text{ form a basis of } \pi^{-1}(x)$$

for each $x \in \mathfrak{U}$.

Assume $\{s_1, \cdots, s_m\}$ is a frame of E over $\mathfrak{U}$. Then it allows us to define a local trivialization of E over $\mathfrak{U}$

$$\varphi : \pi^{-1}(\mathfrak{U}) \longrightarrow \mathfrak{U} \times \mathbb{R}^m$$

by defining for any point $e \in E$, $\pi(e) = x \in \mathfrak{U}$, via

$$\varphi(e) = \{x\} \times (a_1, \cdots, a_m)$$

where $e = a_1 s_1(x) + \cdots + a_m s_m(x)$, $a_i \in \mathbb{R}$. Conversely, any local trivialization of E over $\mathfrak{U}$ gives rise to a frame of E over $\mathfrak{U}$, essentially by the same formula above. Therefore a frame of E may not exist globally on B (unless E is trivial), but always exist locally.

Example 4.1.9 (Tangent bundle)**.** Let X be a smooth manifold of dim $= n$. We can define the **tangent bundle** which has rank n as follows. Set-theoretically, TX is the union

$$TX = \bigcup_{p \in X} T_p X$$

where $T_p X$ is the space of tangent vectors at $p \in X$.

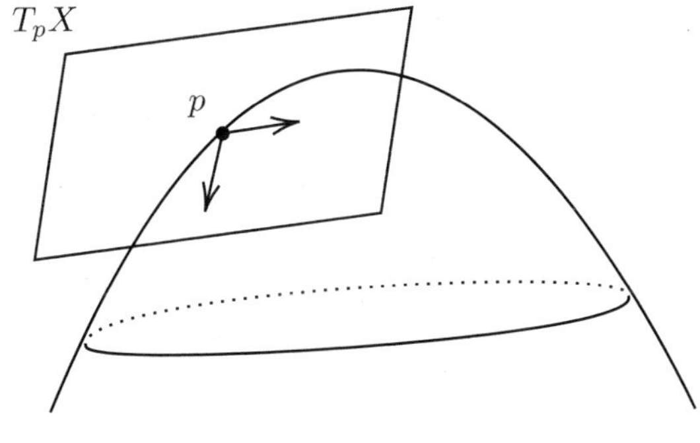

Local trivializations of TX can be constructed with the help of local coordinates. Let $\{x^1, \cdots, x^n\}$ be a local coordinate system on an open subset $\mathfrak{U} \subset X$. Then we have a frame of TX over $\mathfrak{U}$ by the local vector fields

$$\frac{\partial}{\partial x^1}, \cdots, \frac{\partial}{\partial x^n} \qquad \text{on} \quad \mathfrak{U}.$$

A section V of TX over $\mathfrak{U}$ is the same as a vector field on $\mathfrak{U}$, which can be expanded via the frame as

$$V = \sum_{i=1}^{n} V^i(x) \frac{\partial}{\partial x^i}, \qquad x \in \mathfrak{U}.$$

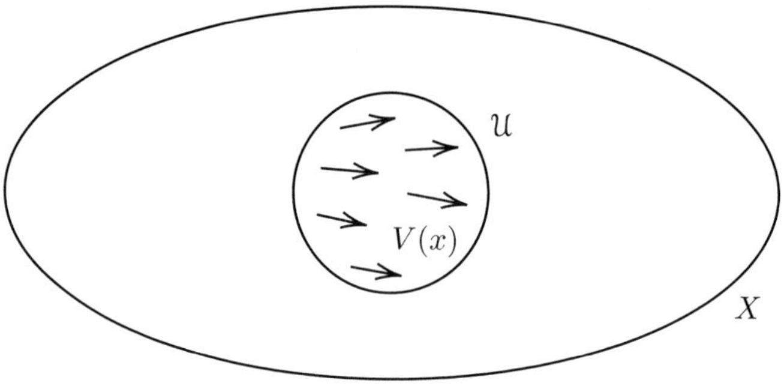

If $\{y^1, \cdots, y^n\}$ is another choice of local coordinates on $\mathfrak{U}'$, then we have the coordinate transformation

$$(x^1, \cdots, x^n) \longmapsto (y^1, \cdots, y^n) \qquad \text{on } \mathfrak{U} \cap \mathfrak{U}'.$$

It gives rise to a linear transformation of the frame via the chain rule

$$\frac{\partial}{\partial x^i} = \sum_{j=1}^{n} \frac{\partial y^j}{\partial x^i} \frac{\partial}{\partial y^j}.$$

In particular, the matrix $\left(\frac{\partial y^j}{\partial x^i}\right)$ is precisely the transition map between two local trivializations of TX. We can expand the same vector field V in both coordinates on the intersection $\mathfrak{U} \cap \mathfrak{U}'$

$$V = \sum_{i=1}^{n} V^i(x) \frac{\partial}{\partial x^i} \qquad \text{in } \{x^i\} \text{ coordinates}$$

$$V = \sum_{j=1}^{n} \widetilde{V}^j(y) \frac{\partial}{\partial y^j} \qquad \text{in } \{y^j\} \text{ coordinates}$$

Then their coefficients are related via the transformation rule of the frame by

$$\widetilde{V}^j(y(x)) = \sum_{i=1}^{n} V^i(x) \frac{\partial y^j}{\partial x^i}.$$

Example 4.1.10 (TS^2). Consider the unit sphere $S^2 \subset \mathbb{R}^3$. Let $p \in S^2$. Then T_pS^2 can be identified with vectors in $\mathbb{R}^3$ orthogonal to p.

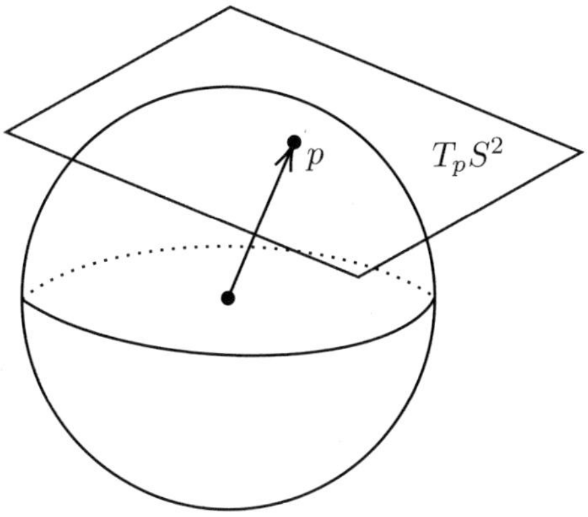

The bundle space of TS^2 can be identified as

$$\begin{array}{ccc} (\vec{r}, \vec{\xi}) & \in & TS^2 = \left\{ (\vec{r}, \vec{\xi}) \in \mathbb{R}^3 \times \mathbb{R}^3 \,|\, |\vec{r}| = 1, \vec{r} \cdot \vec{\xi} = 0 \right\} \\ \big\downarrow & & \big\downarrow \pi \\ \vec{r} & \in & S^2 \end{array}$$

TS^2 is a nontrivial vector bundle: there can not exist a global frame on S^2. In fact, Hairy Ball Theorem says that any global vector field on S^2 must vanish at some point, thus can not be part of a frame everywhere on S^2.

4.1.3 Principal Bundle

Definition 4.1.11. A **principal G-bundle** is a fiber bundle with fiber $F = G$ being a Lie group G. The structure group is a subgroup of G and its action on F is given by the left multiplication

$$\begin{aligned} G \times F(= G) &\longrightarrow F(= G) \\ (g, h) &\longmapsto g \cdot h \end{aligned}$$

We denote a principal G-bundle P over B as

$$\begin{array}{ccc} G & \longrightarrow & P \\ & & \big\downarrow \pi \\ & & B \end{array}$$

The fact that G admits both left and right G-action allows us to define a fiberwise right G-action on P

$$\begin{aligned} P \times G &\longrightarrow P \\ (p, g) &\longmapsto p \cdot g \end{aligned}$$

as follows. Let $b \in B$ and φ_α be a local trivialization. Then φ_α gives a diffeomorphism

$$\varphi_\alpha : \pi^{-1}(b) \xrightarrow{\sim} G.$$

Then for any point $p \in \pi^{-1}(b)$, we define

$$p \cdot g := \varphi_\alpha^{-1}(\varphi_\alpha(p) \cdot g)$$

under the above identification. This does not depend on the choice of the local trivialization. Assume we have another trivialization φ_β which gives a different diffeomorphism

$$\varphi_\beta : \pi^{-1}(b) \xrightarrow{\sim} G.$$

Then the transition function

$$\varphi_{\alpha\beta} : G \xrightarrow{\varphi_\beta^{-1}} \pi^{-1}(b) \xrightarrow{\varphi_\alpha} G$$

is a left multiplication by some element $t \in G$

$$\varphi_{\alpha\beta}(g) = t \cdot g.$$

Since left multiplications commute with right multiplications,

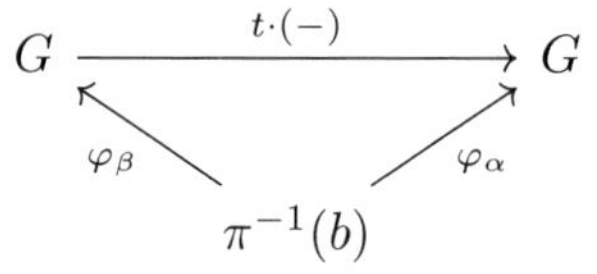

$$\varphi_{\alpha\beta}(\varphi_\beta(p) \cdot g) = t \cdot (\varphi_\beta(p) \cdot g) = (t \cdot \varphi_\beta(p)) \cdot g = \varphi_{\alpha\beta}(\varphi_\beta(p)) \cdot g = \varphi_\alpha(p) \cdot g,$$

which implies

$$\varphi_\alpha^{-1}(\varphi_\alpha(p) \cdot g) = \varphi_\beta^{-1}(\varphi_\beta(p) \cdot g).$$

This says that the right action $p \cdot g$ is well-defined and does not depend on the choice of local trivializations.

Note that the right G-action on itself is transitive and free. Therefore each fiber $\pi^{-1}(b)$ is a single G-orbit. The fibration

$$\pi : P \longrightarrow B$$

can be viewed as the quotient map by the right G-action, and

$$B = P/G$$

can be identified as the orbit space.

Example 4.1.12 (Hopf fibration)**.** The Hopf fibration

$$\begin{array}{ccc} S^1 & \longrightarrow & S^3 \\ & & \downarrow \\ & & S^2 \end{array}$$

can be realized as a principal S^1-bundle. Let us identify

$$S^1 = \{e^{i\theta}\} = \{\lambda \in \mathbb{C} | \, |\lambda| = 1\},$$

$$S^3 = \{(z_1, z_2) \in \mathbb{C}^2 \,|\, |z_1|^2 + |z_2|^2 = 1\}.$$

We define a right S^1-action (Since S^1 is an abelian group, left and right actions are essentially the same) on S^3

$$\begin{aligned} S^3 \times S^1 &\longrightarrow S^3 \\ \left((z_1, z_2), e^{i\theta}\right) &\longmapsto (z_1 e^{i\theta}, z_2 e^{i\theta}) \end{aligned}$$

This action is free. The orbit space

$$S^3/S^1 \simeq \mathbb{C}P^1 = S^2$$

is precisely $\mathbb{C}P^1 = S^2$. This tells that the Hopf fibration is a principal S^1-bundle.

To emphasize the role of a group, we will use $U(1)$ for the group of unitary complex numbers

$$U(1) = \{z \in \mathbb{C} \,|\, |z| = 1\}.$$

This is an abelian Lie group. As we will see, electromagnetism is about the geometry of $U(1)$-principal bundles, hence usually called abelian gauge theory. In general, Yang-Mills theory generalizes Maxwell theory to principal G-bundles, hence is about non-abelian gauge theory.

4.2 $U(1)$-connection and Parallel Transport

We discuss the notion of connection on principal $U(1)$-bundles. Such construction exists on any principal G-bundles, with a bit more care on the non-abelian nature of G. We work on $U(1)$-bundles toward explaining Maxwell theory, and illustrate the basic geometric ideas.

Let $\pi : P \to B$ be a principal $U(1)$-bundle

$$\begin{array}{ccc} U(1) & \longrightarrow & P \\ & & \downarrow \\ & & B \end{array}$$

Geometrically, this is a family of circles

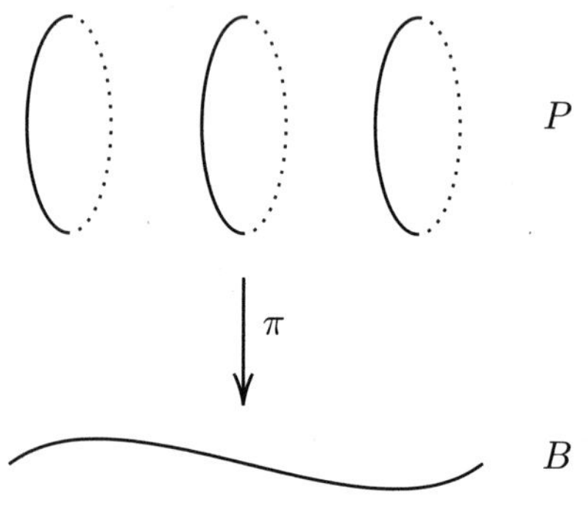

We already know that points of a fiber of P are related to each other via the (right) $U(1)$-action. The essential geometric idea underlying the notion of connection is to be able to transport points of a fiber of P to another fiber along a path in B, so different fibers can communicate and compare with each other.

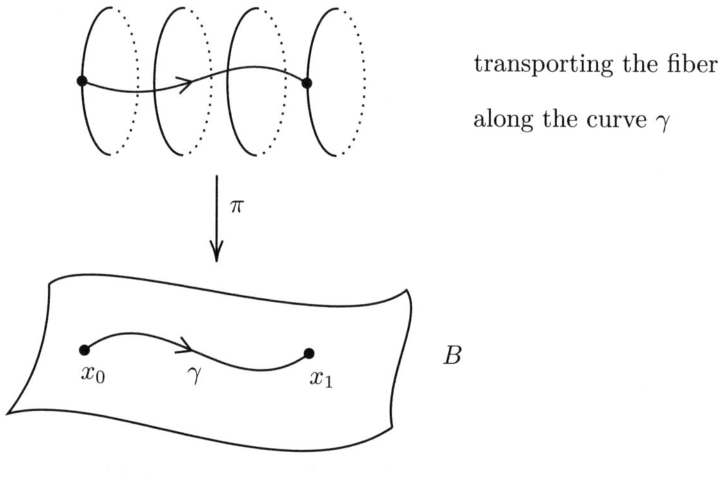

In other words, we need a notion to help lifting a path from the base B to P horizontally.

4.2.1 Vertical Vector Field

The $U(1)$-action on P defines a vector field on P along the fiber direction via its infinitesimal transformation. Let us parametrize $U(1)$ by

$$U(1) = \left\{ e^{i\theta} \right\}.$$

Given a point $x \in P$ and $e^{i\theta} \in U(1)$, the transformed point will be simply denoted by $x \cdot e^{i\theta} \in P$. This defines a curve γ_x

$$\gamma_x : (-\varepsilon, \varepsilon) \longrightarrow P$$
$$\theta \longmapsto x \cdot e^{i\theta}$$

with $\gamma_x(0) = x$. Its derivative at $\theta = 0$ gives a tangent vector

$$\gamma_x'(0) = \frac{d}{d\theta}\bigg|_{\theta=0} \gamma_x(\theta) \in T_x P.$$

This construction applies to all points of P, and gives rise to a vector field on P, denoted by V_θ. Explicitly,

$$V_\theta(x) := \frac{d}{d\theta}\bigg|_{\theta=0} (x \cdot e^{i\theta}) \qquad \text{for } x \in P.$$

The vector field V_θ points along the fiber direction of P, or in formula this means

$$\pi_*(V_\theta) = 0.$$

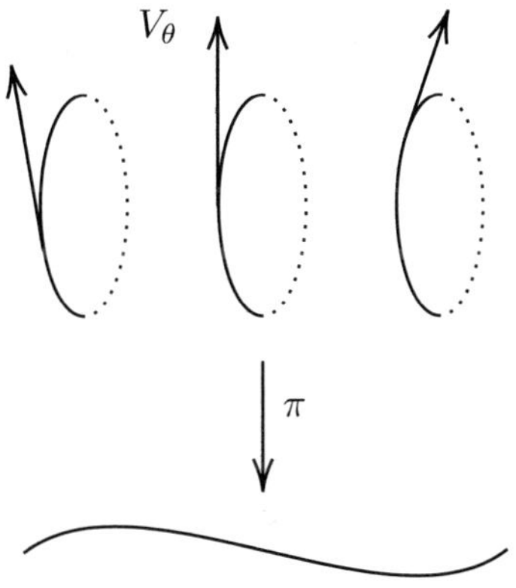

Moreover, if you follow the flow of the vector field V_θ, then it will circle around the fiber and come back to the start point after time 2π.

Remark 4.2.1. In general for a principal G-bundle P, every element of the Lie algebra $\mathfrak{g}$ of G leads to a vector field on P via the infinitesimal right G-action. In this case we have a map

$$\mathfrak{g} \longrightarrow \mathrm{Vect}(P).$$

When $G = U(1)$, $\mathfrak{g} = \mathbb{R}$, it gives a single vector field V_θ as discussed above.

4.2.2 Connection 1-form

Definition 4.2.2. A **connection** on a principal $U(1)$-bundle $\pi : P \to B$ is a 1-form $\mathcal{A} \in \Omega^1(P)$ on P satisfying the following

① $\iota_{V_\theta}\mathcal{A} = 1$ as a function on P. Here ι_{V_θ} is the interior product (contraction) with respect to the vector field V_θ.
② $\mathcal{A}$ is invariant under the $U(1)$-action on P.

Condition ② says that for any $e^{i\theta} \in U(1)$, let us define a diffeomorphism

$$\begin{aligned} f : P &\longrightarrow P \\ x &\longmapsto x \cdot e^{i\theta} \end{aligned}$$

Then $f^*\mathcal{A} = \mathcal{A}$. Equivalently, this is

$$\mathscr{L}_{V_\theta}\mathcal{A} = 0$$

where $\mathscr{L}_{V_\theta}$ is the Lie derivative with respect to the vector field V_θ.

Example 4.2.3. Assume $P = B \times U(1)$ is trivial. Let $\{x^i\}$ denote local coordinates on B and θ denote the angle coordinate on $U(1)$ as above. Then

$$V_\theta = \frac{\partial}{\partial\theta}.$$

An arbitrary 1-form $\mathcal{A}$ can be expressed in coordinates by

$$\mathcal{A} = \mathcal{A}_\theta(x,\theta)d\theta + \sum_i \mathcal{A}_i(x,\theta)dx^i.$$

We have

$$\begin{cases} \iota_{V_\theta}\mathcal{A} = \mathcal{A}_\theta \\ \mathscr{L}_{V_\theta}\mathcal{A} = (\partial_\theta\mathcal{A}_\theta)d\theta + \sum_i(\partial_\theta\mathcal{A}_i)dx^i \end{cases}$$

Then $\mathcal{A}$ defines a connection if
① $\mathcal{A}_\theta = 1$,
② $\partial_\theta\mathcal{A}_\theta = \partial_\theta\mathcal{A}_i = 0$,
i.e. $\mathcal{A}$ is of the form

$$\mathcal{A} = d\theta + \sum_i \mathcal{A}_i(x)dx^i.$$

For a general $U(1)$-bundle, we can describe the 1-form $\mathcal{A}$ in terms of a local trivialization

$$\begin{array}{ccc} \pi^{-1}(\mathfrak{U}_\alpha) & \xrightarrow{\quad\varphi\quad} & \mathfrak{U}_\alpha \times U(1) \\ & \searrow \quad \swarrow & \\ & \mathfrak{U}_\alpha & \end{array}$$

Then a local coordinate system $\{x^i\}$ on $\mathfrak{U}_\alpha$ and the angle coordinate θ on $U(1)$ define a local coordinate system on $\pi^{-1}(\mathfrak{U}_\alpha)$ via the map φ. In such coordinates, the connection $\mathcal{A}$ will again take the form locally

$$(\varphi^{-1})^*\mathcal{A} = d\theta + \sum_i \mathcal{A}_i(x)dx^i$$

for some functions $\mathcal{A}_i(x)$ on $\mathfrak{U}_\alpha$.

4.2.3 Horizontal Vector

A connection 1-form $\mathcal{A}$ allows us to lift a tangent vector from the base manifold to the total space "Horizontally".

Definition 4.2.4. Let $\mathcal{A}$ be a connection on the principal $U(1)$-bundle $\pi : P \to B$. Let $q \in P$. A tangent vector $v \in T_qP$ is called a **Horizontal vector** if

$$\iota_v \mathcal{A} = 0.$$

The space of Horizontal vectors at q will be denoted by

$$H_q = \{v \in T_qP \,|\, \iota_v \mathcal{A} = 0\}.$$

We emphasize that H_q depends on the choice of $\mathcal{A}$.

Clearly we have a direct sum decomposition

$$T_qP = H_q \oplus \mathbb{R}V_\theta.$$

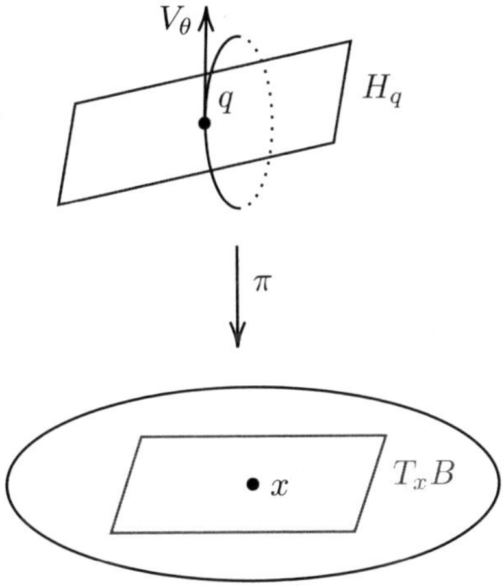

In local coordinates $\{x^i, \theta\}$ from a local trivialization as above, with $\mathcal{A}$ expressed as

$$d\theta + \sum_i \mathcal{A}_i(x)dx^i,$$

the horizontal vectors are spanned by

$$\left\{ \frac{\partial}{\partial x^i} - \mathcal{A}_i(x)\frac{\partial}{\partial \theta} \right\}.$$

It is clear that the push-forward

$$\pi_* : H_q \longrightarrow T_xB, \qquad \text{where } x = \pi(q)$$

is a vector space isomorphism. In coordinates, it sends

$$\pi_* \left(\frac{\partial}{\partial x^i} - \mathcal{A}_i(x)\frac{\partial}{\partial \theta} \right) = \frac{\partial}{\partial x^i}.$$

Definition 4.2.5. Let $q \in P$, $x = \pi(q) \in B$. We define the **horizontal lift** of a tangent vector $v \in T_x B$ at q to be the tangent vector $\widetilde{v} \in H_q$ such that

$$\pi_*(\widetilde{v}) = v.$$

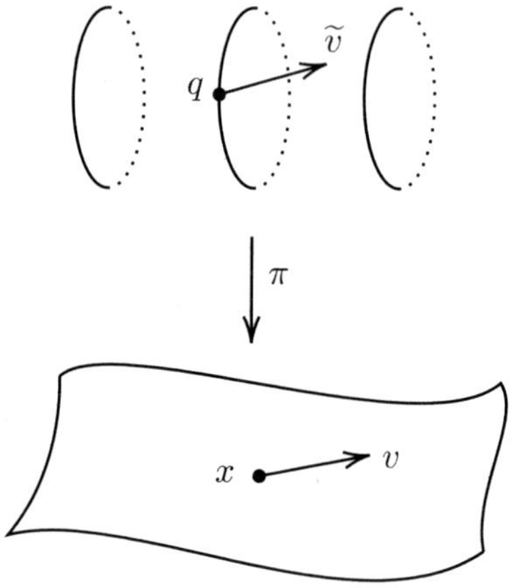

4.2.4 Parallel Transport

Proposition 4.2.6. *Let $\gamma : [0,1] \to B$ be a smooth curve on B from $x_0 = \gamma(0)$ to $x_1 = \gamma(1)$. Let $q_0 \in \pi^{-1}(x_0)$ be any chosen point. Then there exists a unique curve*

$$\widetilde{\gamma} : [0,1] \longrightarrow E$$

with initial condition $\widetilde{\gamma}(0) = q_0$ and such that the tangent vector $\widetilde{\gamma}'(t)$ is the horizontal lift of the tangent vector $\gamma'(t)$ for any $t \in [0,1]$.

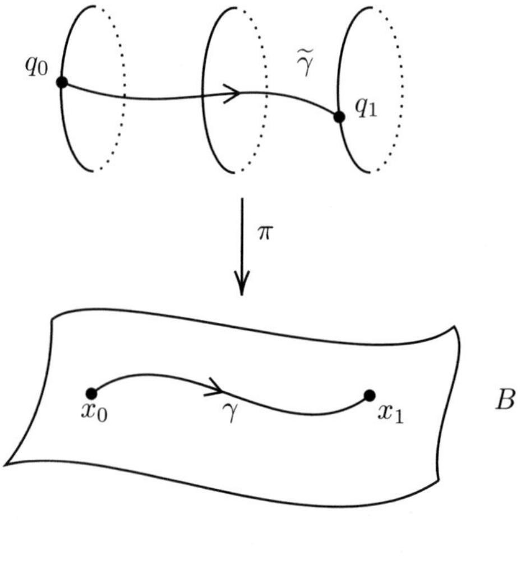

Intuitively, the curve $\widetilde{\gamma}$ is obtained by following the direction of the lifting of that of γ. In local coordinates as above, if $\gamma(t)$ is described by

$$x^i(t) = \gamma^i(t),$$

then $\widetilde{\gamma}(t)$ is given by

$$\widetilde{\gamma}(t) = \big(x^i(t) = \gamma^i(t), \theta(t)\big)$$

where $\theta(t)$ solves the equation (Exercise: show this)

$$\frac{d\theta}{dt} = -\sum_i \left(\frac{d\gamma^i}{dt}\right) \mathcal{A}_i(\gamma(t)).$$

The existence and uniqueness follow from the standard theory in ordinary differential equations.

Definition 4.2.7. Given a curve $\gamma : [0,1] \to B$, with $x_0 = \gamma(0)$ and $x_1 = \gamma(1)$, we define the **parallel transport**

$$T_\gamma : \pi^{-1}(x_0) \longrightarrow \pi^{-1}(x_1)$$

by

$$T_\gamma(q_0) = \widetilde{\gamma}(1)$$

where $\widetilde{\gamma}$ is the horizontal lift of γ with initial condition $\widetilde{\gamma}(0) = q_0$ as in the previous proposition.

Proposition 4.2.8. *$T_\gamma : \pi^{-1}(x_0) \to \pi^{-1}(x_1)$ is $U(1)$-equivariant, i.e.*

$$T_\gamma(q_0 \cdot g) = T_\gamma(q_0) \cdot g, \qquad \forall q_0 \in \pi^{-1}(x_0), g \in U(1).$$

Proof: The fact that $\mathcal{A}$ is $U(1)$-invariant implies that horizontal vectors are preserved under the $U(1)$-action. Assume $\widetilde{\gamma}$ is a horizontal lift of γ. Then for any $g \in U(1)$, the new curve

$$\widetilde{\gamma}_g(t) = \widetilde{\gamma}(t) \cdot g$$

is also a horizontal lifting of γ. If $\widetilde{\gamma}(0) = q_0$, then the initial point of $\widetilde{\gamma}_g$ is

$$\widetilde{\gamma}_g(0) = \widetilde{\gamma}(0) \cdot g = q_0 \cdot g.$$

It follows that

$$T_\gamma(q_0 \cdot g) = \widetilde{\gamma}_g(1) = \widetilde{\gamma}(1) \cdot g = T_\gamma(q_0) \cdot g. \qquad \square$$

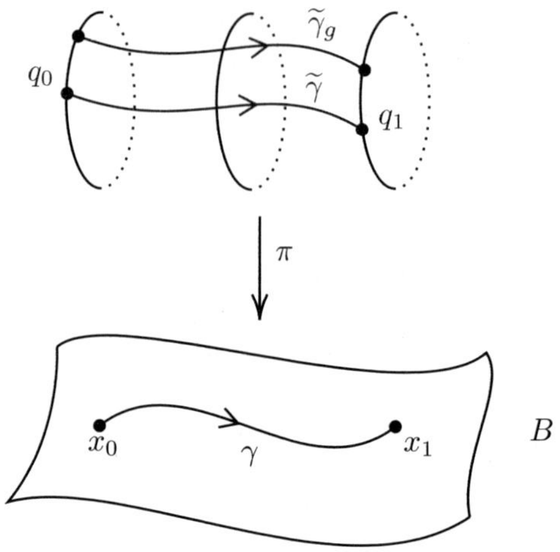

Now we consider the case when γ is a loop, i.e. $\gamma(0) = \gamma(1) = x_0$. Then for any $q_0 \in \pi^{-1}(x_0)$, we have $T_\gamma(q_0) \in \pi^{-1}(x_0)$. Therefore

$$T_\gamma(q_0) = q_0 \cdot g$$

for a unique $g \in U(1)$. This g is called the **holonomy** of γ and q_0, denoted by

$$\mathrm{Hol}_\gamma(q_0).$$

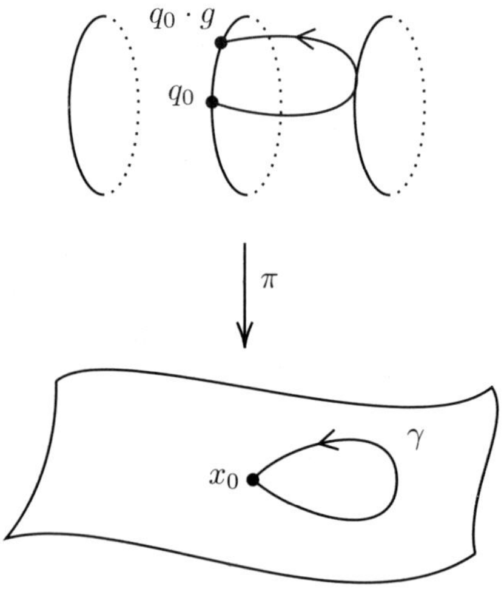

In the $U(1)$-case which is an abelian group, $\mathrm{Hol}_\gamma(q_0)$ does not depend on the choice of q_0 (Exercise: show this) and we will simply write $\mathrm{Hol}_\gamma \in U(1)$.

4.3 Curvature and Chern Class

Let $\mathcal{A}$ be a connection 1-form on the principal $U(1)$-bundle $\pi : P \to B$. We have seen that for any loop $\gamma : [0,1] \to B$ with $\gamma(0) = \gamma(1) = x_0$, it defines a holonomy

$$\mathrm{Hol}_\gamma \in U(1)$$

describing the parallel transport action from the fiber $\pi^{-1}(x_0)$ to itself.

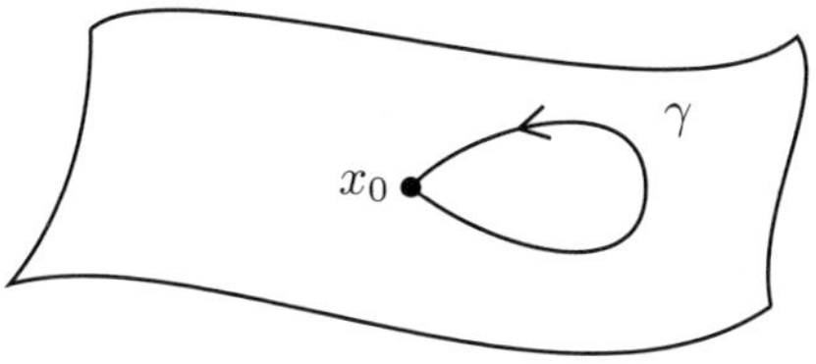

$\mathrm{Hol}_\gamma \in U(1)$

The nontriviality of such holonomies indicate nontriviality of the $U(1)$-bundle via certain twist or curving. We will make this precise in this section. The relevant geometric notion is called the **curvature**.

4.3.1 Curvature 2-form

Definition 4.3.1. Let $\pi : P \to B$ be a principal $U(1)$-bundle, and $\mathcal{A}$ be a connection 1-form. We define its **curvature** by the 2-form

$$F_{\mathcal{A}} = d\mathcal{A}.$$

Since $\mathcal{A}$ is a 1-form on the total space P, the curvature $F_{\mathcal{A}}$ is firstly defined as a 2-form on P. However, $F_{\mathcal{A}}$ is can be viewed in fact as a 2-form on B, which does not depend on the fiber direction. To see this, let us choose a local trivialization over an open $\mathcal{U} \subset B$

$$\begin{array}{ccc} \pi^{-1}(\mathcal{U}) & \longrightarrow & \mathcal{U} \times U(1) \\ \downarrow & \swarrow & \\ \mathcal{U} & & \end{array}$$

with local coordinates $\{x_i\}$ on the base $\mathcal{U}$ and angle coordinate θ on the fiber $U(1)$. The connection $\mathcal{A}$ will take the form

$$\mathcal{A} = d\theta + \sum_i \mathcal{A}_i(x) dx^i.$$

The curvature is then given by

$$F = d\mathcal{A} = \sum_{i,j} \partial_i \mathcal{A}_j dx^i \wedge dx^j = \frac{1}{2} \sum_{i,j} (\partial_i \mathcal{A}_j - \partial_j \mathcal{A}_i) dx^i \wedge dx^j$$

which is clearly a 2-form on the base.

Note that this expression is independent of the choice of local trivialization. In fact, suppose we have a different local trivialization

$$
\begin{array}{ccc}
 & \pi^{-1}(\mathcal{U}) & \\
\swarrow & & \searrow \\
\mathcal{U} \times U(1) & \xrightarrow{\quad\varphi\quad} & \mathcal{U} \times U(1) \\
\cup\!\!\!\!\!\raise2pt\hbox{$\cdot$} & & \cup\!\!\!\!\!\raise2pt\hbox{$\cdot$} \\
(x, e^{i\theta}) & \longmapsto & (x, e^{i\widetilde{\theta}})
\end{array}
$$

Then the transition map

$$(x, e^{i\theta}) \longmapsto (\tilde{x} = x, e^{i\widetilde{\theta}})$$

is given by a family of $U(1)$ -action on the fiber

$$e^{i\theta} = e^{i\phi(x)} e^{i\widetilde{\theta}}$$

for some map

$$e^{i\phi(x)} : \mathcal{U} \longrightarrow U(1).$$

In other words, the coordinate transformation is

$$\begin{cases} x^i = \tilde{x}^i \\ \theta = \widetilde{\theta} + \phi(x) \qquad \mod 2\pi \end{cases}$$

for some function $\phi(x)$. Then in $(\tilde{x}, \widetilde{\theta})$-coordinate,

$$\mathcal{A} = d\theta + \sum_i \mathcal{A}_i(x) dx^i = d\widetilde{\theta} + d\phi(x) + \sum_i \mathcal{A}_i(x) dx^i.$$

The curvature

$$F = d\mathcal{A} = d\left(d\widetilde{\theta} + d\phi(x) + \sum_i \mathcal{A}_i(x) dx^i \right) = \sum_{i,j} (\partial_i \mathcal{A}_j)\, dx^i \wedge dx^j$$

is independent on how to choose the fiber coordinate as expected.

We will treat the curvature $F = d\mathcal{A}$ as a 2-form on B in the subsequent discussions, despite the fact that it is defined in terms of data on P. A more precise statement is that F equals to the pull-back of a 2-form on B via the map π. In fact, the pull-back of forms

$$\pi^* : \Omega^\bullet(B) \to \Omega^\bullet(P)$$

for a fiber bundle is an injective map, and identifies forms on B as a subspace of forms on P (such forms are called *basic forms* in fiber bundle geometry). Then the claim is that the curvature F lies in the image of this map. This explains the above local calculation.

Remark 4.3.2. The curvature takes the simple form $F = d\mathcal{A}$ since $U(1)$ is an abelian Lie group. In general, a connection on a principal G-bundle is a $\mathfrak{g}$-valued 1-form $\mathcal{A} \in \Omega^1(P, \mathfrak{g})$. The curvature will take the form

$$F = d\mathcal{A} + \frac{1}{2}[\mathcal{A}, \mathcal{A}]$$

where $[-,-]$ denotes the Lie bracket on the Lie algebra $\mathfrak{g}$ of G. In the abelian $U(1)$-case, $[-,-] = 0$. In the G-bundle case, such defined curvature form will again descend to the base B.

4.3.2 Holonomy and Curvature

Now we explain the relation between the holonomy and curvature. Let γ be a loop on B which bounds a surface Σ. Here γ could be piece-wise smooth, and Σ may not be a disk but could have non-trivial topology.

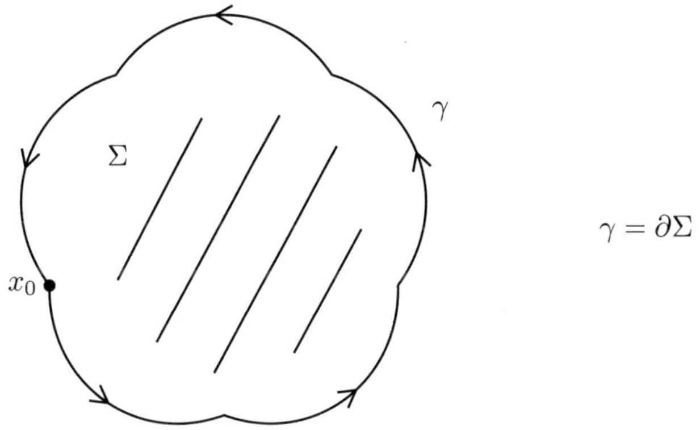

Theorem 4.3.3. *The holonomy of γ is given by the integration of curvature via*

$$\mathrm{Hol}_\gamma = e^{-i\int_\Sigma F} \in U(1).$$

Proof: We triangulate Σ into small pieces as indicated

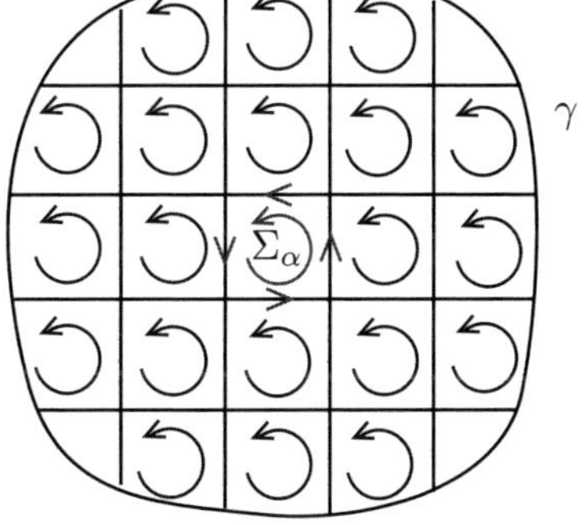

For each small Σ_α, let $\gamma_\alpha = \partial\Sigma_\alpha$ be its boundary loop. For any path in the interior, it is part of two different small loops with opposite direction. The corresponding parallel transport will cancel each other. It follows that

$$\mathrm{Hol}_\gamma = \prod_\alpha \mathrm{Hol}_{\gamma_\alpha}.$$

On the other hand,

$$\int_\Sigma F = \sum_\alpha \int_{\Sigma_\alpha} F.$$

Therefore we only need to prove

$$\mathrm{Hol}_{\gamma_\alpha} = e^{-i\int_{\Sigma_\alpha} F}$$

for each small Σ_α. Thus we assume Σ is small enough and lies inside an open $\mathcal{U} \subset B$ with a local trivialization

$$\begin{array}{ccc} \pi^{-1}(\mathcal{U}) & \longrightarrow & \mathcal{U} \times U(1) \\ & \searrow \quad \swarrow & \\ & \mathcal{U} & \end{array}$$

Let $\{x^i\}$ denote local coordinates on $\mathcal{U}$, and θ denote the angle coordinate on the fiber. The connection 1-form is of the form

$$d\theta + \alpha \qquad \text{where} \qquad \alpha = \sum_i \mathcal{A}_i(x)dx^i.$$

Let $\{x^i(t)\}$ parametrize the curve γ in B. Its lifting $\widetilde{\gamma}$ is parametrized by $\{x^i(t), \theta(t)\}$ where $\theta(t)$ satisfies the flow equation (Horizontal condition)

$$\frac{d\theta}{dt} + \sum_i \mathcal{A}_i(x(t))\frac{dx^i(t)}{dt} = 0$$

which is solved in integral form as

$$\theta(t) = \theta(0) - \int_0^t \gamma^*\alpha, \qquad \text{where} \qquad \gamma : [0,1] \to \mathcal{U}.$$

By definition, the holonomy of γ is

$$\mathrm{Hol}_\gamma = e^{i(\theta(1)-\theta(0))} = e^{-i\int_0^1 \gamma^*\alpha} = e^{-i\int_\gamma \alpha} = e^{-i\int_\Sigma d\alpha} = e^{-i\int_\Sigma F}.$$

This proves the theorem. □

4.3.3 Chern Class

The curvature 2-form F is clearly a closed form on B:

$$dF = 0.$$

It may not be an exact form on B (although it is written as $F = d\mathcal{A}$, but $\mathcal{A}$ is not a 1-form on B but a 1-form on P). Thus F defines a de Rham cohomology class on B which captures topological information of the bundle.

Let us first understand how the curvature depends on the choice of the connection. Assume $\mathcal{A}, \widetilde{\mathcal{A}}$ are two connection 1-forms. Let

$$\alpha = \widetilde{\mathcal{A}} - \mathcal{A}$$

which is a 1-form on P. We claim that α is a pull-back of a 1-form form B. In fact, in local coordinates of a local trivialization, we can write

$$\mathcal{A} = d\theta + \sum_i \mathcal{A}_i(x)dx^i$$

$$\widetilde{\mathcal{A}} = d\theta + \sum_i \widetilde{\mathcal{A}}_i(x)dx^i$$

Then $\alpha = \widetilde{\mathcal{A}} - \mathcal{A} = \sum_i \left(\widetilde{\mathcal{A}}_i(x) - \mathcal{A}_i(x)\right) dx^i$ which clearly depends only on the base. So we will write

$$\alpha \in \Omega^1(B).$$

Remark 4.3.4. The above computation says that the space of connections on P is an affine space

$$\{\text{connections on P}\} = \mathcal{A}_0 + \Omega^1(B)$$

for any specific chosen connection $\mathcal{A}_0$.

Now the curvatures of different connections are related by

$$F_{\widetilde{\mathcal{A}}} = F_{\mathcal{A}} + d\alpha,$$

i.e. differs by an exact 1-form on B. In particular, the de Rham cohomology class

$$[F] \in H^2(B)$$

depends only on the bundle P, but not on the choice of the connection.

Definition 4.3.5. The (first) **Chern class** of the $U(1)$-principal bundle $\pi : P \to B$ is the de Rham cohomology class

$$c_1(P) := \left[\frac{1}{2\pi}F\right] \in H^2(B).$$

One important property of $c_1(P)$ is that it is an integral class. More precisely, we have

Theorem 4.3.6. *Let Σ be any closed surface on B without boundary. Then*

$$\int_\Sigma c_1(P) = \frac{1}{2\pi}\int_\sigma F \in \mathbb{Z}$$

is an integer.

Proof: We can treat Σ as having the boundary with a trivial constant loop γ. Clearly we have $\mathrm{Hol}_\gamma = 1$. By Theorem 4.3.3

$$e^{-i\int_\Sigma F} = \mathrm{Hol}_\gamma = 1 \qquad \Longrightarrow \qquad \int_\Sigma F \in 2\pi\mathbb{Z}.$$

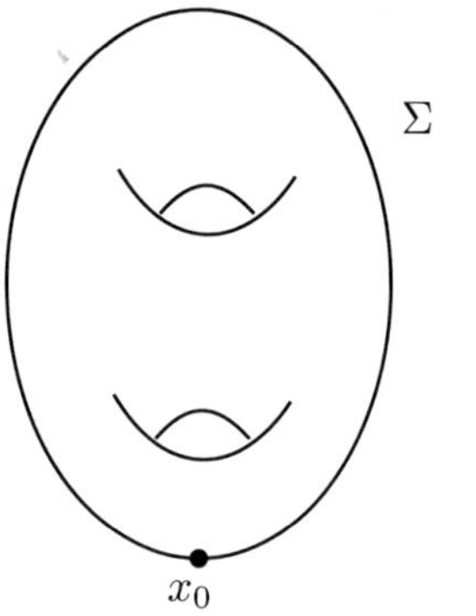

γ: constant loop mapped to x_0

□

Example 4.3.7. If the $U(1)$-bundle $P = B \times U(1)$ is trivial, then

$$c_1(P) = 0 \in H^2(B).$$

In fact, we can choose the connection globally expressed as

$$\mathcal{A} = d\theta.$$

Then $F = d\mathcal{A} = 0$.

Conversely, if we find some closed surface Σ without boundary such that

$$\int_\Sigma F \neq 0,$$

then such bundle can not be trivial. Dirac monopole is such an example (see Example 4.4.5).

4.4 Local Gauge and Transition

We give a concrete description of a connection 1-form in terms of local trivializations.

4.4.1 Local Gauge 1-form via Trivialization

Let

$$\begin{array}{ccc} U(1) & \longrightarrow & P \\ & & \downarrow \pi \\ & & B \end{array}$$

be a principal $U(1)$-bundle. We first give an equivalent description of local trivializations in terms of local sections.

Proposition 4.4.1. *Let $\mathfrak{U} \subset B$. Then there is a one-to-one correspondence*

$$\{\textit{local trivializations of } P \textit{ over } \mathfrak{U}\} \overset{1:1}{\longleftrightarrow} \{\textit{local sections of } P \textit{ over } \mathfrak{U}\}$$

$$\Downarrow \qquad\qquad\qquad \Downarrow$$

$$\varphi : \pi^{-1}(U) \to \mathfrak{U} \times U(1) \longleftrightarrow \sigma \in \Gamma(\mathfrak{U}, P)$$

Proof: Let $\sigma \in \Gamma(\mathfrak{U}, P)$. Then it defines a diffeomorphism

$$\begin{aligned} \varphi^{-1} : \mathfrak{U} \times U(1) &\longrightarrow \pi^{-1}(\mathfrak{U}) \\ (x, g) &\longmapsto \sigma(x) \cdot g \end{aligned}$$

Its inverse defines a local trivialization φ.

Conversely, given a local trivialization φ, it defines a local section σ by

$$\sigma(x) = \varphi^{-1}(x, 1)$$

where $1 \in U(1)$ is the identity element. This establishes the correspondence.

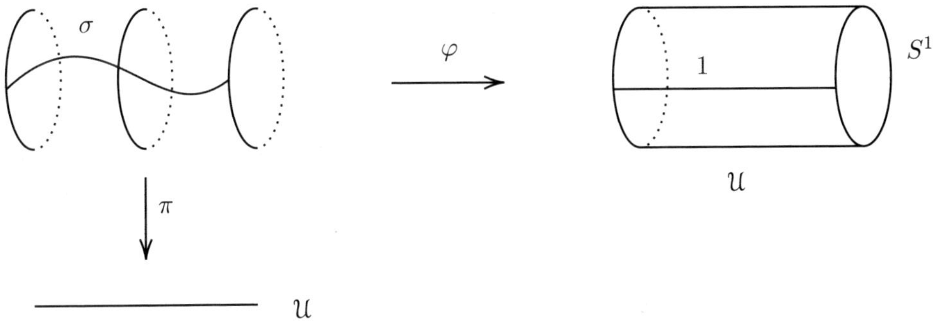

□

Let $\mathcal{A}$ be a connection 1-form on P. We aim at an explicit description of $\mathcal{A}$ locally. Let

$$\varphi : \pi^{-1}(\mathfrak{U}) \longrightarrow \mathfrak{U} \times U(1)$$

be a trivialization over an open $\mathfrak{U} \subset B$. It corresponds to a section $\sigma \in \Gamma(\mathfrak{U}, P)$ via

$$\sigma(x) = \varphi^{-1}(x, 1)$$

as described in the previous proposition. Let $\{x^i\}$ be local coordinates on $\mathfrak{U}$, and θ be the angle coordinate on $\mathfrak{U}(1)$. As we have described before, the connection $\mathcal{A}$ in the local trivialization will take the form

$$d\theta + \sum_i \mathcal{A}_i(x) dx^i.$$

Precisely, this means that

$$\mathcal{A}|_{\pi^{-1}(\mathfrak{U})} = \varphi^* \left(d\theta + \sum_i \mathcal{A}_i(x) dx^i \right)$$

for some $\mathcal{A}_i(x)$. To describe the meaning of $\mathcal{A}_i(x)$, consider the section $\sigma : \mathfrak{U} \to P$. Let

$$A_\sigma := \sigma^* \mathcal{A}$$

which is a 1-form on $\mathfrak{U}$. To compute A_σ, consider the composition

$$\mathfrak{U} \xrightarrow{\sigma} \pi^{-1}(\mathfrak{U}) \xrightarrow{\varphi} \mathfrak{U} \times U(1).$$

We have

$$A_\sigma = \sigma^* \mathcal{A} = \sigma^* \varphi^* \left(d\theta + \sum_i \mathcal{A}_i(x) dx^i \right) = (\varphi \circ \sigma)^* \left(d\theta + \sum_i \mathcal{A}_i(x) dx^i \right).$$

Since

$$\begin{aligned} \varphi \circ \sigma : \mathfrak{U} &\longrightarrow \mathfrak{U} \times U(1) \\ x &\longmapsto (x, 1) \end{aligned}$$

It follows that

$$A_\sigma = \sum_i \mathcal{A}_i(x) dx^i.$$

In other words, the information of $\mathcal{A}_i(x)$ in the chosen local trivialization is precisely the pull-back of $\mathcal{A}$ to the base via the corresponding local section.

We next describe how A_σ depends on the choice of local trivializations. Let

$$\widetilde{\varphi} : \pi^{-1}(\mathfrak{U}) \to \mathfrak{U} \times U(1)$$

be another local trivialization on $\mathfrak{U}$, which corresponds to another section $\widetilde{\sigma} \in \Gamma(\mathfrak{U}, P)$. Then it determines a map

$$e^{i\alpha(x)} : \mathfrak{U} \longrightarrow U(1)$$

such that

$$\widetilde{\sigma}(x) = \sigma(x) \cdot e^{i\alpha(x)}, \qquad \forall x \in \mathfrak{U}.$$

It relates the two trivializations by

$$\sigma(x) \cdot g = \widetilde{\sigma}(x) \cdot e^{-i\alpha(x)} g.$$

$$\begin{array}{ccc} & \pi^{-1}(\mathfrak{U}) & \\ \swarrow^{\varphi} & & \searrow^{\widetilde{\varphi}} \\ \mathfrak{U} \times U(1) & \xrightarrow{\widetilde{\varphi} \circ \varphi^{-1}} & \mathfrak{U} \times U(1) \\ \Cup & & \Cup \\ (x, g) & \longmapsto & (x, e^{-i\alpha(x)} g) \end{array}$$

Let

$$\begin{cases} A_\sigma = \sigma^* \mathcal{A} = \sum_i \mathcal{A}_i(x) dx^i \\ \widetilde{A}_\sigma = \widetilde{\sigma}^* \mathcal{A} = \sum_i \widetilde{\mathcal{A}}_i(x) dx^i \end{cases}$$

be local descriptions of $\mathcal{A}$ with respect to the corresponding trivialization. By construction, under the diffeomorphism

$$\mathfrak{U} \times U(1) \xrightarrow{\widetilde{\varphi} \circ \varphi^{-1}} \mathfrak{U} \times U(1),$$

we have

$$(\widetilde{\varphi} \circ \varphi^{-1})^* \left(d\theta + \sum_i \widetilde{\mathcal{A}}_i(x) dx^i \right) = d\theta + \sum_i \mathcal{A}_i(x) dx^i.$$

On the other hand, since

$$\widetilde{\varphi} \circ \varphi^{-1} : (x, e^{i\theta}) \longmapsto \left(x, e^{-i\alpha(x)} e^{i\theta} \right),$$

we have explicitly

$$\begin{aligned} (\widetilde{\varphi} \circ \varphi^{-1})^* \left(d\theta + \sum_i \widetilde{\mathcal{A}}_i(x) dx^i \right) &= d\left(\theta - \alpha(x)\right) + \sum_i \widetilde{\mathcal{A}}_i(x) dx^i \\ &= d\theta - d\alpha(x) + \sum_i \widetilde{\mathcal{A}}_i(x) dx^i. \end{aligned}$$

Thus we find

$$\sum_i \widetilde{\mathcal{A}}_i(x) dx^i = \sum_i \mathcal{A}_i(x) dx^i + d\alpha(x).$$

To summarize, we have proved the following

Proposition 4.4.2. *Let $\mathcal{A}$ be a connection 1-form on P. Let $\sigma, \widetilde{\sigma} \in \Gamma(\mathfrak{U}, P)$ be two sections over an open $\mathfrak{U} \subset B$, which are related by*

$$\widetilde{\sigma} = \sigma \cdot e^{i\alpha}$$

for some $e^{i\alpha} : \mathfrak{U} \to U(1)$. Then the two local descriptions $A_\sigma = \sigma^* \mathcal{A}$ and $A_{\widetilde{\sigma}} = \widetilde{\sigma}^* \mathcal{A}$ are related by

$$A_{\widetilde{\sigma}} = A_\sigma + d\alpha.$$

Definition 4.4.3. A_σ is called the **local gauge 1-form** of $\mathcal{A}$ with respect to the trivialization σ. The above change of A_σ via different local trivializations is called "**local gauge transformations**".

4.4.2 Connection via Transition Functions

Let $\{\mathfrak{U}_\alpha\}$ be an open cover of B, with local trivializations

$$\varphi_\alpha : \pi^{-1}(\mathfrak{U}_\alpha) \longrightarrow \mathfrak{U}_\alpha \times U(1)$$

and transition function

$$\varphi_{\beta\alpha} : \mathfrak{U}_\alpha \cap \mathfrak{U}_\beta \longrightarrow U(1)$$

such that

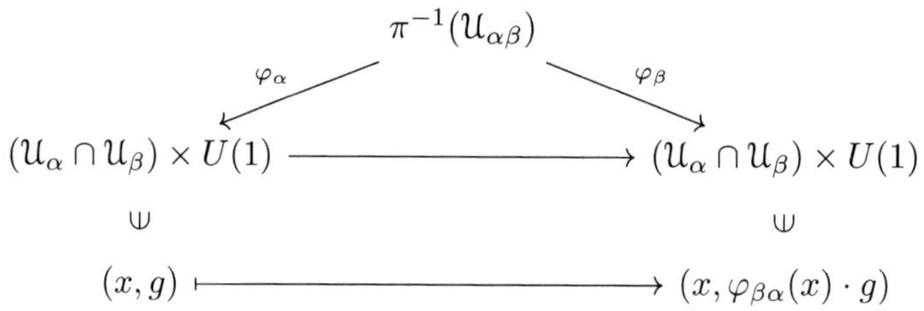

The collection of transition functions $\{\varphi_{\alpha\beta}\}$ clearly satisfies the following

① $\varphi_{\alpha\beta} = \varphi_{\beta\alpha}^{-1}$,

② $\varphi_{\alpha\gamma}(x)\varphi_{\gamma\beta}(x)\varphi_{\beta\alpha}(x) = 1$, $\forall x \in \mathfrak{U}_\alpha \cap \mathfrak{U}_\beta \cap \mathfrak{U}_\gamma$.

Condition ② is also called the **cocycle condition**.

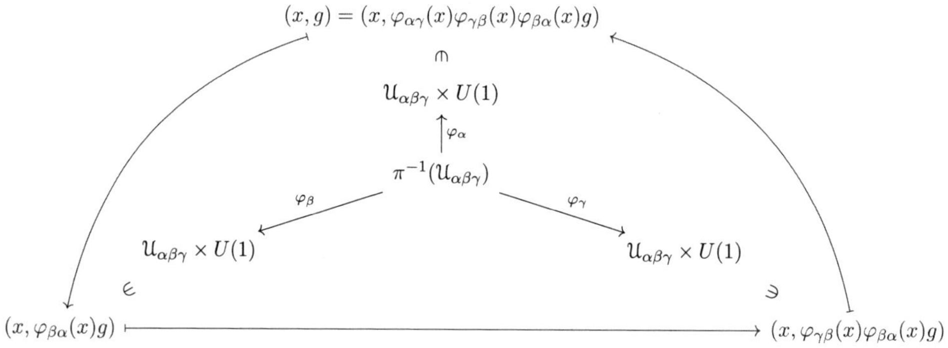

Here $\mathfrak{U}_{\alpha\beta\gamma} = \mathfrak{U}_\alpha \cap \mathfrak{U}_\beta \cap \mathfrak{U}_\gamma$.

In fact, any collection of maps

$$\varphi_{\beta\alpha} : \mathfrak{U}_\alpha \cap \mathfrak{U}_\beta \longrightarrow U(1)$$

satisfying condition ① and ② defines a principal $U(1)$-bundle P by the quotient

$$P = \left(\coprod_\alpha \mathfrak{U}_\alpha \times U(1) \right) \Big/ \sim$$

where the equivalence relation $\sim$ identifies

$$\begin{array}{ccc} (x,g) & \sim & (x,\varphi_{\beta\alpha}(x)g) \\ \cap\!\!\!\!\text{|} & & \cap\!\!\!\!\text{|} \\ \mathfrak{U}_\alpha \times U(1) & & \mathfrak{U}_\beta \times U(1) \end{array}$$

for $x \in \mathfrak{U}_\alpha \cap \mathfrak{U}_\beta$. We leave the details to the reader to check that this is indeed a principal $U(1)$-bundle.

Now let us reformulate the data of local trivializations $\{\varphi_\alpha\}$ and transition functions $\{\varphi_{\alpha\beta}\}$. By the proposition above, each φ_α corresponds to a local section $\sigma_\alpha : \mathfrak{U} \to \pi^{-1}(\mathfrak{U})$ by

$$\sigma_\alpha(x) = \varphi_\alpha^{-1}(x, 1).$$

Then for $x \in \mathfrak{U}_\alpha \cap \mathfrak{U}_\beta$, we have

$$\sigma_\beta(x) = \varphi_\alpha^{-1} \circ \varphi_\alpha \circ \varphi_\beta^{-1}(x,1) = \varphi_\alpha^{-1}(x, \varphi_{\alpha\beta}(x)) = \varphi_\alpha^{-1}(x,1)\varphi_{\alpha\beta}(x) = \sigma_\alpha(x)\varphi_{\alpha\beta}(x),$$

i.e. we have

$$\sigma_\beta = \sigma_\alpha \varphi_{\alpha\beta} \qquad \text{on } \mathfrak{U}_\alpha \cap \mathfrak{U}_\beta.$$

Let us fix a trivialization of P described by local sections $\sigma_\alpha : \mathfrak{U}_\alpha \to P$ on each $\mathfrak{U}_\alpha$ and transition functions $\{\varphi_{\alpha\beta} : \mathfrak{U}_\alpha \cap \mathfrak{U}_\beta \to U(1)\}$ as above. Let $\mathcal{A}$ be a connection 1-form on P. We can use σ_α to define a local gauge 1-form A_α on $\mathfrak{U}_\alpha$ via pull-back

$$A_\alpha := \sigma_\alpha^*(\mathcal{A}) \in \Omega^1(\mathfrak{U}_\alpha).$$

On the intersection $\mathfrak{U}_\alpha \cap \mathfrak{U}_\beta$, we have

$$\sigma_\beta = \sigma_\alpha \varphi_{\alpha\beta} \qquad \text{on} \quad \mathfrak{U}_\alpha \cap \mathfrak{U}_\beta.$$

By Proposition 4.4.2, we have

$$A_\beta = A_a + \frac{1}{i}\varphi_{\alpha\beta}^{-1} d\varphi_{\alpha\beta} \tag{$*$}$$

Equivalently, if we write $\varphi_{\alpha\beta} = e^{i\chi_{\alpha\beta}}$, then

$$A_\beta = A_\alpha + d\chi_{\alpha\beta}.$$

Note that the curvature is $F_{\mathcal{A}} = dA_\alpha$, and it glues $F_\alpha = F_\beta$ on $\mathfrak{U}_\alpha \cap \mathfrak{U}_\beta$ to define a global 2-form on B. This is consistent with our previous discussion. Conversely, local 1-forms $\{A_\alpha \in \Omega^1(\mathfrak{U})\}$ which are related by equation $(*)$ on intersections give rise to a connection 1-form $\mathcal{A}$ on P, by running back the above process. We have now proved the following

Proposition 4.4.4. *Let $\pi : P \to B$ be a principal $U(1)$-bundle with chosen local trivializations $\{\sigma_\alpha \in \Gamma(\mathfrak{U}_\alpha, P)\}$ on an open cover $\{\mathfrak{U}_\alpha\}$ and transition functions $\{\varphi_{\alpha\beta} : \mathfrak{U}_\alpha \cap \mathfrak{U}_\beta \to U(1)\}$ such that $\sigma_\beta = \sigma_\alpha\varphi_{\alpha\beta}$ on $\mathfrak{U}_\alpha \cap \mathfrak{U}_\beta$. Then there is a one-to-one correspondence*

$$
\begin{array}{ccc}
\{\textit{Connection 1-form } \mathcal{A} \textit{ on } P\} & \xleftrightarrow{1:1} & \left\{ \begin{array}{l} \textit{A collection of 1-form } A_\alpha \textit{ on each } \mathfrak{U}_\alpha \textit{ such that} \\ A_\beta = A_\alpha + \frac{1}{i}\varphi_{\alpha\beta}^{-1} d\varphi_{\alpha\beta} \textit{ on each } \mathfrak{U}_\alpha \cap \mathfrak{U}_\beta \end{array} \right\} \\
\Psi & & \Psi \\
\mathcal{A} & \longmapsto & \{A_\alpha = \sigma_\alpha^* \mathcal{A}\}
\end{array}
$$

Example 4.4.5 (Dirac Monopole). Consider the unit sphere $S^2 \subset \mathbb{R}^3$ with spherical coordinates (θ, φ).

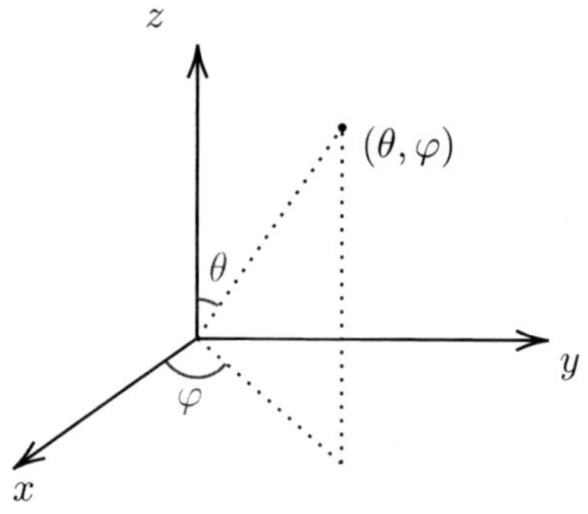

We consider a cover of S^2 by

$$\mathfrak{U}^+ = S^2 - \{(0,0,-1)\}$$

$$\mathfrak{U}^- = S^2 - \{(0,0,1)\}$$

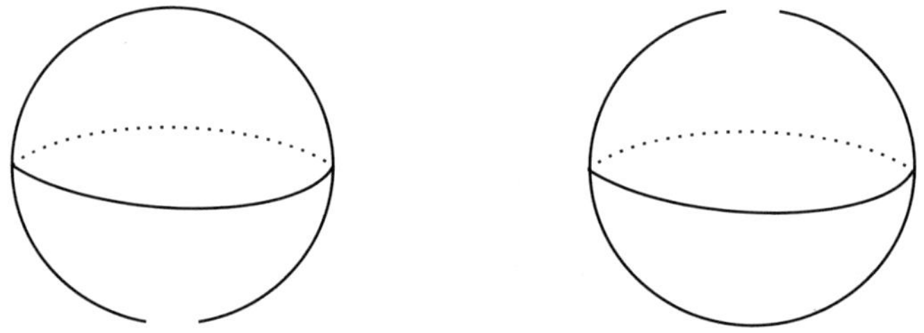

$\mathfrak{U}_+$ $\mathfrak{U}_-$

Each $\mathfrak{U}_\pm$ is contractible, hence the fiber bundle can be trivialized on $\mathfrak{U}_\pm$.

Consider the following 1-form

$$A_\pm = \pm\frac{n}{2}(1 \mp \cos\theta)\, d\varphi \qquad \text{on} \qquad \mathfrak{U}_\pm.$$

Here n is a constant to be determined. On $\mathfrak{U}_+ \cap \mathfrak{U}_-$,

$$A_+ - A_- = n d\varphi.$$

If we ask $\{A_+, A_-\}$ to define a connection 1-form of some principal $U(1)$-bundle, then the transition function on $\mathfrak{U}_+ \cap \mathfrak{U}_-$ is

$$\begin{aligned} \mathfrak{U}_+ \cap \mathfrak{U}_- &\longrightarrow U(1) \\ (\theta, \varphi) &\longmapsto e^{in\varphi} \end{aligned}$$

Since φ is defined modulo 2π, this map is well-defined if and only if $n \in \mathbb{Z}$ is an integer.

In other words, for each integer $n \in \mathbb{Z}$, the transition function

$$\begin{aligned} e^{i\chi} : \mathfrak{U}_+ \cap \mathfrak{U}_- &\longrightarrow U(1) \\ (\theta, \varphi) &\longmapsto e^{in\varphi} \end{aligned}$$

defines a principal $U(1)$-bundle P_n on S^2, and the collection $\{A_\pm = \pm\frac{n}{2}(1 \mp \cos\theta)\, d\varphi\}$ defines a connection on P_n. Then the curvature form is

$$F = dA_+ \,(= dA_-) = \frac{n}{2} \sin\theta \, d\theta \wedge d\varphi.$$

It is direct to compute

$$\frac{1}{2\pi} \int_{S^2} F = n.$$

Hence the $U(1)$-bundle P_n is nontrivial for $n \neq 0$. Moreover, these $U(1)$-bundle $\{P_n\}$'s are all topologically different since they have different Chern classes (Theorem 4.3.6).

4.5 Gauge Transformation

We describe the group of gauge transformations on principal $U(1)$-bundles. Such transformations give rise to equivalent data both in physics and in geometry.

4.5.1 Gauge Transformation

Let $\pi : P \to B$ be a principal $U(1)$-bundle. We define an **automorphism** of P to be a diffeomorphism

$$f : P \longrightarrow P$$

such that

① f is fiberwise, i.e. $\pi \circ f = \pi$

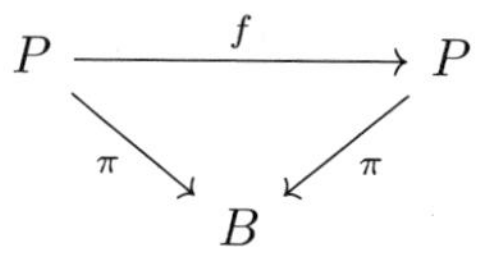

② f is $U(1)$-equivariant, i.e.

$$f(q \cdot g) = f(q) \cdot g, \qquad \forall q \in P, g \in U(1).$$

We denote $\mathrm{Aut}(P)$ to be the group of all automorphisms of the principal $U(1)$-bundle P.

For each point $x \in B$, f defines a $U(1)$-equivariant map

$$f : \pi^{-1}(x) \longrightarrow \pi^{-1}(x).$$

Since $U(1)$ is abelian, this is simply given by a right action by an element of $U(1)$. It follows that such f can be equivalently described by

$$\widetilde{f} : B \longrightarrow U(1)$$

such that

$$f(q) = q \cdot \widetilde{f}(x), \qquad \forall q \in \pi^{-1}(x).$$

Thus we have proved

Proposition 4.5.1. *There is a group isomorphism*

$$\mathrm{Aut}(P) = \mathrm{Map}(B, U(1)).$$

Here given $\widetilde{f}_1, \widetilde{f}_2 \in \mathrm{Map}(B, U(1))$, *their group product is*

$$\widetilde{f}_1 \cdot \widetilde{f}_2(x) := \widetilde{f}_1(x) \cdot \widetilde{f}_2(x), \qquad \forall x \in B.$$

Let $\mathcal{A}$ be a connection 1-form on P. Let $f \in \mathrm{Aut}(P)$ be an automorphism of P. We can use f to pull-back $\mathcal{A}$ to get another 1-form

$$f^*\mathcal{A} \in \Omega^1(P).$$

We claim that $f^*\mathcal{A}$ is also a connection. In fact, since f is $U(1)$-equivariant, for any $q \in P$,

$$\left.\frac{\partial}{\partial\theta}\right|_{\theta=0} \left(f(qe^{i\theta})\right) = \left.\frac{\partial}{\partial\theta}\right|_{\theta=0} \left(f(q)e^{i\theta}\right),$$

which says

$$f_* V_\theta = V_\theta.$$

It follows that

① $\iota_{V_\theta} f^*\mathcal{A} = f^*(\iota_{V_\theta}\mathcal{A}) = f^*(1) = 1$

② $\mathscr{L}_{V_\theta} f^*\mathcal{A} = f^*(\mathscr{L}_{V_\theta}\mathcal{A}) = f^*(0) = 0.$

We can also describe this explicitly in local coordinates via a local trivializations, with base coordinates $\{x^i\}$ and $U(1)$ angle coordinate θ. Then $\mathcal{A}$ can be written as

$$\mathcal{A} = d\theta + \sum_i A_i(x)dx^i.$$

The automorphism f can be expressed by a map

$$\widetilde{f} : B \longrightarrow U(1)$$

which in local coordinates is

$$\widetilde{f}(x) = e^{i\phi(x)}.$$

Then

$$f(x, e^{i\theta}) = (x, e^{i\theta}e^{i\phi(x)}),$$

so

$$f^*\mathcal{A} = d(\theta + \phi(x)) + \sum_i \mathcal{A}_i(x)dx^i = d\theta + \sum_i \mathcal{A}_i(x)dx^i + d\phi(x)$$

which still takes the form of a connection. This proves the claim that $f^*\mathcal{A}$ is also a connection.

Proposition 4.5.2. *Let $\mathcal{A}$ be a connection 1-form. Let $f \in \operatorname{Aut}(P)$ which corresponds to $\widetilde{f} = e^{i\phi} \in \operatorname{Map}(B, U(1))$ as above. Then*

$$f^*\mathcal{A} = \mathcal{A} + \pi^* d\phi.$$

Proof: This follows from the above local computation. □

Remark 4.5.3. ϕ is defined modulo 2π, but $d\phi$ is a well-defined 1-form.

Definition 4.5.4. Let $\operatorname{Conn}(P)$ denote the space of connections on P. The natural action

$$\operatorname{Aut}(P) \curvearrowright \operatorname{Conn}(P)$$

is called **global gauge transformations**, or simply **gauge transformations**.

Proposition 4.5.5. *The curvature 2-form of the principal $U(1)$-bundle P is invariant under gauge transformations.*

Proof: Let $\mathcal{A} \in \operatorname{Conn}(P)$ and $f \in \operatorname{Aut}(P)$. Then

$$F_{f^*\mathcal{A}} = d(f^*\mathcal{A}) = d\mathcal{A} + d(d\phi) = d\mathcal{A} = F_{\mathcal{A}}.$$

□

Remark 4.5.6. In general for principal G-bundle, the curvature 2-form will transform via group conjugation under gauge transformations. When $G = U(1)$ is abelian, group conjugation is trivial.

Therefore the curvature of $U(1)$-bundle can be viewed as a map

$$F : \operatorname{Conn}(P)/\operatorname{Aut}(P) \longrightarrow \Omega^2(B).$$

In summary, any two different choices of connection differ by a 1-form on B: $\widetilde{\mathcal{A}} = \mathcal{A} + \pi^*\alpha$. The curvatures differ by the exact form $d\alpha$

$$F_{\widetilde{\mathcal{A}}} = F_{\mathcal{A}} + d\alpha.$$

Gauge transformations lead to change of the connection by $\alpha = d\phi$, hence leaving F invariant.

4.5.2 Local vs. Global Gauge Transformations

The formula of local and global gauge transformations look very similar. We clarify these two notions to avoid possible confusions. In short,

- Local gauge transformation is about the different expression of the same connection via different choices of local trivializations.
- Global gauge transformation is about the change from one connection to another, so linking different connections.

However, these two notions are closely related when the bundle P is trivial, where we have a trivialization over the full base B. Indeed, assume now P is trivial, then we have an identification (Proposition 4.4.1)

$$\{\text{trivialization of } P \text{ over } B\} = \Gamma(B, P).$$

Let $\sigma_1, \sigma_2 \in \Gamma(B, P)$ be two trivializations. Then there exists a unique map

$$\widetilde{f} : B \longrightarrow U(1)$$

such that

$$\sigma_2(x) = \sigma_1(x)\widetilde{f}(x), \qquad \forall x \in B.$$

Let $f \in \mathrm{Aut}(P)$ be the automorphism corresponding to $\widetilde{f}$.

The above relations are equivalently described as

$$\sigma_2 = f \circ \sigma_1.$$

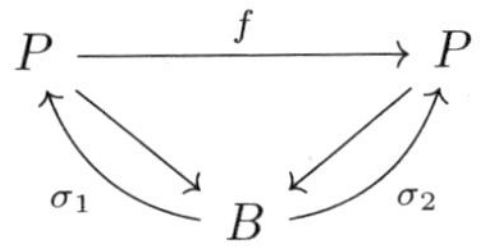

In particular, $\mathrm{Aut}(P)$ acts on $\Gamma(B, P)$ freely and transitively in the case when P is trivial.

Let $\mathcal{A}$ be a connection on a trivial $U(1)$-bundle P over B. Let $\sigma \in \Gamma(B, P)$ be a global section which gives a trivialization of P. Let $f \in \mathrm{Aut}(P)$ be an automorphism of P which corresponds to a map

$$\begin{aligned} \widetilde{f} : B &\longrightarrow U(1) \\ x &\longmapsto e^{i\phi(x)} \end{aligned}$$

① f gives a (global) gauge transformation

$$\mathcal{A} \longmapsto \widetilde{\mathcal{A}} = f^*\mathcal{A} = \mathcal{A} + \pi^* d\phi.$$

In the local form described by the same trivialization σ, we have

$$\begin{cases} A_\sigma = \sigma^*\mathcal{A} \\ \widetilde{A}_\sigma = \sigma^*(f^*\mathcal{A}) = A_\sigma + d\phi \end{cases}$$

② f gives a different trivialization by the section

$$\widetilde{\sigma} = f \circ \sigma \qquad \in \Gamma(B, P).$$

These two sections σ and $\widetilde{\sigma}$ give different local descriptions of the same connection $\mathcal{A}$ by

$$\begin{cases} A_\sigma = \sigma^* \mathcal{A} \\ A_{\widetilde{\sigma}} = \widetilde{\sigma}^* \mathcal{A} = A_\sigma + d\phi \end{cases}$$

Although the transformation formula in ① and ② look similarly, they are different in nature. ① is about gauge transformations, which changes one connection to another connection. ② is about local gauge transformation, where the connection is fixed but expressed on the base with different trivializations. In the case when P is trivial, we have a (noncanonical) identification

$$\Gamma(B, P) \simeq \mathrm{Aut}(P) \qquad \text{if } P \text{ is trivial.}$$

In general, they are different and we hope it will not cause further confusion.

Now we discuss the case when P is nontrivial in general. Let us fix an open cover $\{\mathfrak{U}_\alpha\}$ of B and local trivialization on each $\mathfrak{U}_\alpha$ by $\sigma_\alpha \in \Gamma(\mathfrak{U}_\alpha, P)$. Let $\mathcal{A}$ be a connection 1-form on P. Then we can express $\mathcal{A}$ locally via the collection $\{\sigma_\alpha\}$ by

$$\{A_\alpha = \sigma_\alpha^* \mathcal{A}\}\,.$$

Let $f \in \mathrm{Aut}(P)$ be an automorphism of P, which corresponds to a map

$$\widetilde{f} = e^{i\phi} : B \longrightarrow U(1).$$

The gauge transformation of $\mathcal{A}$ via f is

$$\mathcal{A} \longmapsto f^* \mathcal{A}.$$

We can express the gauge transformation locally via the fixed trivialization $\{\sigma_\alpha\}$

$$\{A_\alpha\} \longmapsto \{(f^*\mathcal{A})_\alpha\}\,.$$

Then locally on each $\mathfrak{U}_\alpha$, we have

$$(f^*\mathcal{A})_\alpha = (f \circ \sigma_\alpha)^* \mathcal{A} = A_\alpha + d\phi.$$

Thus a gauge transformation is equivalently described with respect to the fixed trivialization $\{\sigma_\alpha\}$ by a transformation of collections

$$\{A_\alpha\} \overset{f}{\longmapsto} \{A_\alpha + d\phi\}\,, \qquad \text{where } d\phi \text{ is the same expression in all } \alpha$$

for $\widetilde{f} = e^{i\phi}$ as above. Again, the gauge transformation preserves the curvature form

$$\{F_\alpha = dA_\alpha\} \overset{f}{\longmapsto} \{(f^*F)_\alpha = dA_\alpha + d(d\phi) = dA_\alpha\}\,.$$

4.6 Maxwell Theory as $U(1)$-gauge Theory

We are now ready to explain electromagnetism in terms of $U(1)$-gauge theory.

4.6.1 Potential as Gauge 1-form

Let us now consider the base manifold being the spacetime

$$B = \mathbb{R}^{3,1}.$$

Let P be a principal $U(1)$-bundle on $\mathbb{R}^{3,1}$. Since $\mathbb{R}^{3,1}$ is contractible, P is a trivial bundle. We will fix once for all a global section σ of P that gives a global trivialization

$$P \xrightarrow{\sim} \mathbb{R}^{3,1} \times U(1).$$

Any connection 1-form $\mathcal{A}$ on P therefore corresponds equivalently to a gauge 1-form A on B via

$$A_\sigma = \sigma^* \mathcal{A}.$$

Since σ will be fixed, we will simply write the gauge 1-form as $A = A_\sigma$. In spacetime coordinates $\{x, y, z, t\}$ on $\mathbb{R}^{3,1}$,

$$A = A_t dt + A_x dx + A_y dy + A_z dz.$$

The gauge 1-form is identified with potentials in electromagnetism as

$$\begin{cases} \phi = -A_t & \text{scalar potential} \\ \vec{\mathbf{A}} = (A_x, A_y, A_z) & \text{vector potential} \end{cases}$$

Let $e^{i\chi} : \mathbb{R}^{3,1} \to U(1)$. It gives a gauge transformation of the connection 1-form, hence the (local) gauge 1-form, via

$$A \longmapsto A + d\chi.$$

This is the same gauge transformation that we have encountered in electromagnetism.

4.6.2 Electromagnetic Field as Curvature

The curvature 2-form of the connection

$$F = dA$$

collects the components of electric and magnetic fields. It takes the form that we have seen

$$F = \mathbb{E} \wedge dt + *_3 \mathbb{B} = (\mathbf{E}_x dx + \mathbf{E}_y dy + \mathbf{E}_z dz) \wedge dt + (\mathbf{B}_x dy \wedge dz + \mathbf{B}_y dz \wedge dx + \mathbf{B}_z dx \wedge dy).$$

The curvature of the $U(1)$-connection is invariant under gauge transformations.

4.6.3 Maxwell Action

Let J denote the electric charge-current 1-form

$$J = \rho/\varepsilon_0\, dt - \mu_0(\mathbf{j}_x dx + \mathbf{j}_y dy + \mathbf{j}_z dz).$$

Then Maxwell's equations are

$$\begin{cases} dF = 0 \\ d^* F = J \end{cases}$$

Here $d^* = *d*$ is the adjoint of d. These equations can be derived from the action principle as follows.

Definition 4.6.1. The **Maxwell action** $S_M[A]$ is a functional on $\mathrm{Conn}(P)$ defined by

$$S_M[A] = \frac{1}{\mu_0}\int_{\mathbb{R}^{3,1}} \left(\frac{1}{2}F \wedge *F + *J \wedge A\right).$$

If we write it as

$$S_M[A] = \int_{\mathbb{R}^{3,1}} \mathscr{L}(A)\, dx \wedge dy \wedge dz \wedge cdt,$$

where $\mathscr{L}(A)$ is the Lagrangian density, then a direct computation gives

$$\mathscr{L} = \frac{\varepsilon_0}{2}\vec{\mathbf{E}} \cdot \vec{\mathbf{E}} - \frac{1}{2\mu_0}\vec{\mathbf{B}} \cdot \vec{\mathbf{B}} + \left(-\rho\phi + \vec{\mathbf{A}} \cdot \vec{\mathbf{j}}\right).$$

Now we consider the variation of $S_M[A]$ under an arbitrary variation of the connection

$$A \longrightarrow A + \delta A$$

where δA is a 1-form on $\mathbb{R}^{3,1}$. Under this variation,

$$\begin{aligned} \delta S_M &= \frac{1}{\mu_0}\int_{\mathbb{R}^{3,1}} \frac{1}{2}(\delta F) \wedge *F + \frac{1}{2}F \wedge *\delta F + *J \wedge \delta A \\ &= \frac{1}{\mu_0}\int_{\mathbb{R}^{3,1}} (\delta F \wedge *F - \delta A \wedge *J) \\ &= \frac{1}{\mu_0}\int_{\mathbb{R}^{3,1}} (d\delta A \wedge *F - \delta A \wedge *J) \\ &= \frac{1}{\mu_0}\int_{\mathbb{R}^{3,1}} \delta A \wedge (d(*F) - *J). \end{aligned}$$

An extremal point of S_M corresponds to a gauge 1-form A such that

$$\delta S_M = 0 \qquad \text{for arbitrary } \delta A.$$

This is the same as asking

$$d(*F) = *J$$

which is equivalent to $d^* F = J$. On the other hand, the equation $dF = 0$ is already captured by the use of potential. We conclude that

$$\text{Solution of Maxwell's equations} \quad \Longleftrightarrow \quad \text{Critical point of } S_M$$

4.6.4 Gauge Principle and Charge Conservation

Gauge principle asks for invariance of physical quantities under gauge transformations. Let us analyze the Maxwell action under gauge transformations.

Consider a gauge transformation described by

$$e^{i\chi} : \mathbb{R}^{3,1} \longrightarrow U(1).$$

χ can be viewed as a function on $\mathbb{R}^{3,1}$ defined modulo 2π. Since the topology of $\mathbb{R}^{3,1}$ is trivial, we can actually lift χ to be a single-valued function on $\mathbb{R}^{3,1}$, and we assume this.

Under the gauge transformation generated by $e^{i\chi}$, the gauge 1-form transforms as

$$A \longmapsto A + d\chi$$

and the curvature F is invariant. It follows that the Maxwell action transforms as

$$S_M[A + d\chi] = S_M[A] + \frac{1}{\mu_0} \int_{\mathbb{R}^{3,1}} *J \wedge d\chi = S_M[A] + \frac{1}{\mu_0} \int_{\mathbb{R}^{3,1}} d(*J) \wedge \chi.$$

If we require that S_M should be invariant under arbitrary gauge transformations, then we need

$$d(*J) = 0.$$

This is precisely the charge conservation. In other words, charge conservation can be viewed as a direct consequence of gauge principle. Assume charge conservation, then the Maxwell action defines a functional

$$S_M : \mathrm{Conn}(P)/\,\mathrm{Aut}(P) \longrightarrow \mathbb{R}.$$

4.6.5 Magnetic Monopole

Maxwell's equations

$$\begin{cases} dF = 0 \\ d(*F) = *J \end{cases}$$

is based on the assumption (in known experiments) that magnetic monopole does not exist. In theory, the full Maxwell's equations are

$$\begin{cases} dF = *J_m \\ d(*F) = *J_e \end{cases}$$

where J_e is the electric charge-current 1-form, and J_m is the magnetic charge-current 1-form. It exhibits the full electromagnetic duality

$$F \longleftrightarrow *F$$
$$J_m \longleftrightarrow J_e$$

The case without magnetic monopole corresponds to $J_m = 0$. In general, suppose $J_m \neq 0$ but is supported on a subspace $M \subset \mathbb{R}^{3,1}$. Thus

$$J_m = 0 \qquad \text{on} \qquad \mathbb{R}^{3,1} - M$$

and on the region $\mathbb{R}^{3,1} - M$ we have

$$\begin{cases} dF = 0 \\ d(*F) = *J_e \end{cases} \qquad \text{on} \qquad \mathbb{R}^{3,1} - M.$$

Therefore it corresponds to a $U(1)$-gauge theory on $\mathbb{R}^{3,1} - M$. However, the topology of $\mathbb{R}^{3,1} - M$ may no longer be trivial, and there could exist nontrivial principal $U(1)$-bundle on $\mathbb{R}^{3,1} - M$. In this case the connection can not be described by a single gauge 1-form, but need a collection of gauge 1-forms on a covering of $\mathbb{R}^{3,1} - M$.

We describe the example of Dirac magnetic monopole to illustrate this. Consider a magnetic monopole with magnetic charge n sitting at rest at the origin of $\mathbb{R}^3$. Its trajectory in spacetime $\mathbb{R}^{3,1}$ is

$$M = \{0\} \times \mathbb{R}_t \subset \mathbb{R}^3 \times \mathbb{R}_t$$

where $\mathbb{R}_t$ denotes the real line parameterized by time t. In this case we are led to consider $U(1)$-gauge theory on

$$\mathbb{R}^{3,1} - M = \left(\mathbb{R}^3 - \{0\}\right) \times \mathbb{R}_t.$$

We use spherical coordinates (r, θ, φ) to parametrize points in the space.

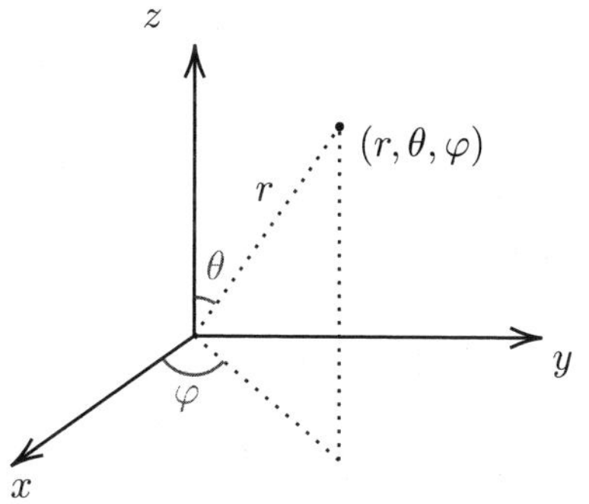

Then $\mathbb{R}^{3,1}$ is parametrized by (r, θ, φ, t) and $\mathbb{R}^{3,1} - M$ corresponds to the locus where $r > 0$.

Now we choose a covering of $\mathbb{R}^{3,1} - M$ by

$$\begin{aligned} \mathcal{U}_+ &= \left(\mathbb{R}^3 \setminus \{(0,0,z) \mid z \leq 0\}\right) \times \mathbb{R}_t \\ \mathcal{U}_- &= \left(\mathbb{R}^3 \setminus \{(0,0,z) \mid z \geq 0\}\right) \times \mathbb{R}_t \end{aligned}$$

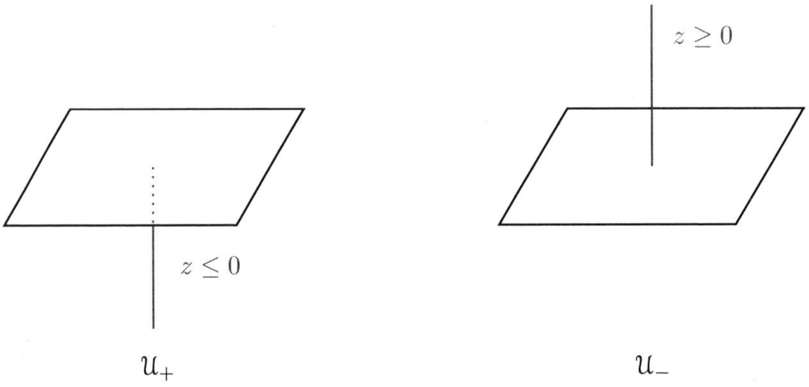

Each $\mathfrak{U}_\pm$ is contractible and so a principal $U(1)$-bundle can be trivialized on the cover $\{\mathfrak{U}_+, \mathfrak{U}_-\}$. The local gauge 1-form of the Dirac magnetic monopole is described in the local trivialization by

$$A_\pm = \frac{n}{2r(z \pm r)}(-ydx + xdy) = \pm\frac{n}{2}(1 \mp \cos\theta)d\varphi \qquad \text{on } \mathfrak{U}_\pm.$$

On the intersection $\mathfrak{U}_+ \cap \mathfrak{U}_-$, we have

$$A_+ - A_- = nd\varphi \qquad \text{on } \mathfrak{U}_+ \cap \mathfrak{U}_-$$

which corresponds to the transition function (assuming $n \in Z$)

$$e^{in\varphi} : \mathfrak{U}_+ \cap \mathfrak{U}_- \longrightarrow U(1).$$

This transition function defines a principal $U(1)$-bundle P_n on $\mathbb{R}^{3,1} - M$ (see Example 4.4.5). The curvature form is

$$F = dA_+ (= dA_-) = \frac{n}{2}\sin\theta d\theta \wedge d\varphi.$$

Let S_r^2 denote a sphere of radius r in $\mathbb{R}^3$. Then

$$\int_{S_r^2} c_1(P_n) = \frac{1}{2\pi}\int_{S_r^2} F = n \in \mathbb{Z}.$$

This describes a magnetic monopole of magnetic charge n. In fact

$$\frac{1}{2\pi}\int_{S^2} F = \frac{1}{2\pi}\int_{S^2} *_3\mathbb{B} = \frac{1}{2\pi}\int_{S^2} d\vec{\sigma}\cdot\vec{\mathbf{B}}$$

which describes the magnetic flux over S^2. It computes the total magnetic charge surrounded by the sphere S_r^2.

4.7 Associated Bundle and Matter Field

At the end, we describe how electromagnetic field is coupled with matter field.

4.7.1 Associated Vector Bundle

Let us start with a general discussion of the construction of vector bundles associated to a principal G-bundle. Let

$$\begin{array}{ccc} G & \longrightarrow & P \\ & & \downarrow{\scriptstyle \pi} \\ & & B \end{array}$$

be a principal G-bundle. Let

$$\rho : G \longrightarrow \mathrm{End}(V)$$

be a representation of G on a vector space V of dimension m. We consider the quotient space

$$P \times_\rho V := P \times V / \sim$$

where the equivalence relation $\sim$ is

$$(p \cdot g, v) \sim (p, g \cdot v), \qquad \forall p \in P, v \in V, g \in G.$$

Here $p \cdot g$ is the right g-action on P, and $g \cdot v$ is the left g-action on V with respect to the representation ρ

$$g \cdot v := \rho(g)(v).$$

It carries a natural projection map

$$\begin{array}{ccc} P \times_\rho V & \ni & (p, v) \\ \downarrow{\scriptstyle \pi_\rho} & & \downarrow \\ B & \ni & \pi(p) \end{array}$$

Proposition 4.7.1. *$\pi_\rho : P \times_\rho V \to B$ defines a vector bundle of* rank $= \dim V$.

Proof: Let

$$\begin{array}{ccc} \pi^{-1}(\mathfrak{U}) & \xrightarrow{\varphi} & \mathfrak{U} \times G \\ & \searrow \quad \swarrow & \\ & \mathfrak{U} & \end{array}$$

be a local trivialization of P over an open $\mathfrak{U} \subset B$. Then it induces

$$\begin{array}{ccccc} \pi^{-1}(\mathfrak{U}) \times_G V & \longrightarrow & (\mathfrak{U} \times G) \times_G V & \longrightarrow & \mathfrak{U} \times V \\ & \searrow & \downarrow & \swarrow & \\ & & \mathfrak{U} & & \end{array}$$

which defines a local trivialization of $P \times_G V$ over $\mathfrak{U}$ with fiber $= V$. It can be checked that the transition functions are linear transformations on V, with the help of the representation map $\rho : G \to \mathrm{End}(V)$. $\square$

Definition 4.7.2. $P \times_\rho V$ is called the **associated vector bundle** of the principal G-bundle P with respect to the G-representation ρ on V.

In the following discussion, we will be mainly discussing principal $U(1)$-bundles. Irreducible unitary representation of $U(1)$ is always on $\mathbb{C}$, with the representation ρ_n classified by

$$\rho_n : e^{i\theta} \longmapsto e^{in\theta}, \qquad n \in \mathbb{Z}.$$

The associated vector bundle is a complex vector bundle of rank $= 1$, i.e. a complex line bundle.

4.7.2 Hermitian Inner Product

Let us fix a principal $U(1)$-bundle $\pi : P \to B$ and the $U(1)$-representation ρ_n on $\mathbb{C}$ as above. Let

$$\mathscr{L}_n = P \times_{\rho_n} \mathbb{C} \xrightarrow{\pi_\rho} B$$

denote the associated complex line bundle.

For any point $x \in B$, the fiber $\pi_{\rho_n}^{-1}(x)$ is a one dimensional complex vector space. We can define a **Hermitian inner product** on each finer $\pi_{\rho_n}^{-1}(x)$ as follows. Let $v \in \pi_{\rho_n}^{-1}(x)$, which is represented by a pair (p, z) for $p \in \pi^{-1}(x), z \in \mathbb{C}$. Then we define its norm by

$$|v|^2 := |z|^2.$$

This is independent of the choice of the representative. In fact, another representative of v is $\left(pe^{-i\theta}, \rho_n(e^{i\theta})z\right) = (pe^{-i\theta}, e^{in\theta}z)$. Then

$$\left|e^{in\theta}z\right|^2 = |z|^2$$

which has the same value.

In general, for a section $\psi \in \Gamma(B, \mathscr{L}_n)$, we can define its norm pointwise

$$|\psi|^2 \in \mathbb{C}^\infty(B).$$

As we will discuss below, the wave function of particle in quantum mechanics will be a section of a complex line bundle $\mathscr{L}$ as above, and $|\psi|^2$ plays the role of probability distribution of the quantum particle.

4.7.3 Covariant Derivative

We know that we can take derivatives on functions. We would like to extend such notion of derivative on sections of vector bundles. We will focus on the case of the complex line bundle $\mathscr{L}_n$ arising from the principal $U(1)$-bundle P and the $U(1)$-representation ρ_n on $\mathbb{C}$ as above. We explain that a connection 1-form $\mathcal{A}$ on P will allow us to define derivatives of sections of all the associated bundles.

Precisely, let $s \in \Gamma(B, \mathscr{L}_n)$ denote a section of $\mathscr{L}_n$ on B, and $V \in \mathrm{Vect}(B)$ be a vector field on B. We will define a notion of **covariant derivative** of s with respect to the vector field V, denoted by $\nabla_V s$, which is again a section $\nabla_V s \in \Gamma(B, \mathscr{L}_n)$ as follows. Let $\mathfrak{U} \subset B$ be an open subset, with a local trivialization of P defined by a local section

$$\sigma \in \Gamma(\mathfrak{U}, P).$$

It allows us to specify a unique representative of $s(x)$ for $x \in \mathfrak{U}$ by

$$s(x) = (\sigma(x), f(x)), \qquad x \in \mathfrak{U}.$$

Here $f(x)$ is a complex valued function on $\mathfrak{U}$. This representation clearly depends on the choice of σ. Now we define the section $\nabla_V s$ whose value on $\mathfrak{U}$ is represented by

$$\nabla_V s = (\sigma, \partial_V f + in\ \iota_V(A_\sigma) f).$$

Here $A_\sigma = \sigma^* \mathcal{A}$ is the local gauge 1-form with respect to the choice σ. Note that since f is a function, $\partial_V f$ is the usual derivative with respect to V. We can write it in a compact form

$$\nabla_V s = (\sigma, \iota_V (df + inA_\sigma f)),$$

where $df + inA_\sigma f$ is a 1-form on $\mathfrak{U}$.

We check that such $\nabla_V s$ is well-defined, i.e. does not depend on the choice of σ. Assume we have another section $\widetilde{\sigma} \in \Gamma(\mathfrak{U}, P)$. Then

$$\widetilde{\sigma}(x) = \sigma(x) e^{i\phi(x)}, \qquad x \in \mathfrak{U},$$

where

$$e^{i\phi(x)} : \mathfrak{U} \longrightarrow U(1).$$

Using $\widetilde{\sigma}$, we can represent the section s as

$$s(x) = \left(\widetilde{\sigma}(x), \widetilde{f}(x)\right).$$

Now since $(\sigma(x), f(x)) \sim (\widetilde{\sigma}(x), \widetilde{f}(x))$, which should represent the same element $s(x)$, we find

$$\widetilde{f}(x) = \rho_n(e^{-i\phi(x)}) f(x) = e^{-in\phi(x)} f(x).$$

On the other hand, the local gauge 1-form $A_{\widetilde{\sigma}}$ is related to A_σ by a local transformation

$$A_{\widetilde{\sigma}} = A_\sigma + d\phi.$$

Now let us compute $\nabla_V s$ with respect to the local section $\widetilde{\sigma}$. It gives

$$\left(\widetilde{\sigma}, \iota_V \left(d\widetilde{f} + inA_{\widetilde{\sigma}} \widetilde{f}\right)\right).$$

We need to check that it defines the same section of $\mathscr{L}_n$ on $\mathfrak{U}$ as

$$(\sigma, \iota_V (df + inA_\sigma f)).$$

Indeed, from

$$d\widetilde{f} + inA_{\widetilde{\sigma}} \widetilde{f} = d\left(e^{-in\phi} f\right) + in\left(A_\sigma + d\phi\right) e^{-in\phi} f = e^{-in\phi}\left(df + inA_\sigma f\right).$$

We see that

$$\begin{aligned}\left(\widetilde{\sigma}, \iota_V\left(d\widetilde{f}+inA_{\widetilde{\sigma}}\widetilde{f}\right)\right) &= \left(\widetilde{\sigma}, e^{-in\phi}\iota_V\left(df+inA_\sigma f\right)\right) \sim \left(\widetilde{\sigma}e^{-i\phi}, \iota_V\left(df+inA_\sigma f\right)\right)\\ &= \left(\sigma, \iota_V\left(df+inA_\sigma f\right)\right).\end{aligned}$$

This shows that $\nabla_V s$ is well-defined and does not depend on the choice of local trivialization.

4.7.4 Matter Wave Function

In our previous study of electromagnetism, the electric and magnetic fields are of central role, while the scalar and vector potentials are introduced as an auxiliary object to help solve Maxwell's equations. However in quantum mechanics, the situation is completely changed and the use of gauge potential is essential. We briefly mention some key constructions.

As we have discussed before, the Maxwell theory of electromagnetism can be viewed as a $U(1)$-gauge theory

connection	$\longleftrightarrow$	potentials
curvature	$\longleftrightarrow$	electromagnetic fields

In quantum mechanics, particles are described by wave functions ψ, where $|\psi|^2$ describes the probability distribution of the quantum particle. In an electromagnetic background, a particle of electric charge n will be described by a section

$$s \in \Gamma(\mathbb{R}^{3,1}, \mathscr{L}_n)$$

of the complex line bundle $\mathscr{L}_n$ associated to the principal $U(1)$-bundle P of Maxwell theory. It seems unnecessary to introduce $U(1)$-bundle here since the topology of $\mathbb{R}^{3,1}$ is trivial. However, when the space-time encodes nontrivial topology, this description will be essential. As we have seen before, the norm $|s|^2$ is still well-defined, with the physical meaning of probability amplitude. With respect to a choice of section σ on $\mathbb{R}^{3,1}$, we can represent the section as

$$s = (\sigma, \psi)$$

for ψ a function on $\mathbb{R}^{3,1}$. Then

$$|s(x)|^2 = |\psi(x)|^2.$$

If we choose another section to represent s, then ψ will undertake a phase rotation

$$\psi \longrightarrow e^{in\psi}$$

but leaving $|\psi|^2$ invariant.

Now the coupling between electromagnetism and matter field s is through the construction of covariant derivative associated to a connection 1-form on P. Let us fix σ as before, and let

$$A = \sigma^* \mathcal{A} = -\phi dt + \mathbf{A}_x dx + \mathbf{A}_y dy + \mathbf{A}_z dz$$

be the corresponding gauge 1-form. Then we have covariant derivatives on s as defined in Section 4.7.3. For example,

$$\nabla_x s = (\sigma, \partial_x \psi + in\mathbf{A}_x \psi)\,.$$

Sometimes we simply write it as

$$\nabla_x \psi = \partial_x \psi + in\mathbf{A}_x \psi.$$

For the Schrödinger equation governing the wave function ψ, we simply need to replace every derivative such as $\partial_x, \partial_y, \partial_z$ by $\nabla_x = \partial_x + in\mathbf{A}_x$, $\nabla_y = \partial_y + in\mathbf{A}_y$, $\nabla_z = \partial_z + in\mathbf{A}_z$. For example, the Hamiltonian $\mathscr{H}$ is

$$\mathscr{H} = -\frac{1}{2m} \sum_i \left(\frac{\partial}{\partial x^i} + in\mathbf{A}_i \right)^2 + V(r).$$

This explains the basic principle on how gauge field is coupled with matter field.

Chapter 5

Electromagnetism and Special Relativity

The theory of special relativity originates from electromagnetism. The work of Hendrik Lorentz through his study of electromagnetism plays a fundamental role. In this chapter, we will study basic aspects of the coherence of electromagnetism with respect to special relativity. We will see in particular how Lorentz transformations preserve the form of Maxwell's equations.

5.1 Lorentz Transformation

5.1.1 Lorentz Group

We will parametrize the spacetime $\mathbb{R}^{3,1}$ via

$$x^\mu = (ct, x, y, z) \qquad \mu = 0, 1, 2, 3.$$

We consider the following Minkowski distance

$$c^2t^2 - x^2 - y^2 - z^2 = \sum_{\mu,\nu} \eta_{\mu\nu} x^\mu x^\nu,$$

where

$$\eta = \begin{pmatrix} 1 & 0 & 0 & 0 \\ 0 & -1 & 0 & 0 \\ 0 & 0 & -1 & 0 \\ 0 & 0 & 0 & -1 \end{pmatrix}$$

is called the **Minkowski metric**. The space $\mathbb{R}^{3,1}$ equipped with the Minkowski metric is called **Minkowski spacetime**.

Definition 5.1.1. A **Lorentz transformation** is a linear transformation $\Lambda : \mathbb{R}^{3,1} \to \mathbb{R}^{3,1}$ preserving the Minkowski distance. The Lorentz group is the group of all Lorentz transformations, denoted by $O(3,1)$.

In coordinates, if

$$\Lambda : x^\mu \longmapsto \widetilde{x}^\mu = \sum_\nu \Lambda^\mu{}_\nu x^\nu,$$

then the 4×4 matrix $\Lambda^\mu{}_\nu$ gives a Lorentz transformation if it obeys

$$\sum_{\mu,\nu} \eta_{\mu\nu} \Lambda^\mu{}_\rho \Lambda^\nu{}_\sigma = \eta_{\rho\sigma}, \qquad \forall \rho, \sigma = 0, 1, 2, 3$$

or equivalently in matrix form

$$\Lambda^T \eta \Lambda = \eta, \qquad \Lambda^T = \text{transpose of } \Lambda.$$

Example 5.1.2. Let $R \in O(3)$ be a 3×3 rotation on $\mathbb{R}^3$. Then $R^T \cdot R = I$. It gives a Lorentz transformation

$$\Lambda = \begin{pmatrix} 1 & 0 & 0 & 0 \\ 0 & & & \\ 0 & & R & \\ 0 & & & \end{pmatrix}$$

which is simply a rotation on (x, y, z) and leaves t invariant. It preserves the Euclidean distance $x^2 + y^2 + z^2$, hence preserves the Minkowski distance $c^2t^2 - x^2 - y^2 - z^2$.

Example 5.1.3 (Lorentz Boost). A Lorentz boost along x-direction with velocity v is the linear transformation

$$\begin{cases} \widetilde{ct} = \dfrac{ct - xv/c}{\sqrt{1 - v^2/c^2}} \\ \widetilde{x} = \dfrac{x - vt}{\sqrt{1 - v^2/c^2}} \\ \widetilde{y} = y \\ \widetilde{z} = z \end{cases}$$

It is direct to check that

$$(\widetilde{ct})^2 - \widetilde{x}^2 - \widetilde{y}^2 - \widetilde{z}^2 = (ct)^2 - x^2 - y^2 - z^2.$$

This corresponds to a Lorentz transformation with matrix form

$$\Lambda = \begin{pmatrix} \gamma & -\gamma v/c & 0 & 0 \\ -\gamma v/c & \gamma & 0 & 0 \\ 0 & 0 & 1 & 0 \\ 0 & 0 & 0 & 1 \end{pmatrix}$$

where $\gamma = \frac{1}{\sqrt{1 - v^2/c^2}}$. We can view this as a hyperbolic rotation in the (x^0, x^1)-plane

$$\Lambda = \begin{pmatrix} \cosh\alpha & -\sinh\alpha & 0 & 0 \\ -\sinh\alpha & \cosh\alpha & 0 & 0 \\ 0 & 0 & 1 & 0 \\ 0 & 0 & 0 & 1 \end{pmatrix}.$$

Similarly, we have Lorentz boosts along y and z directions.

Observe that the relation

$$\Lambda^T \eta \Lambda = \eta$$

implies

$$\det(\Lambda)^2 = 1 \qquad \Longrightarrow \qquad \det(\Lambda) = \pm 1.$$

Definition 5.1.4. The group of **proper Lorentz transformations** is defined as

$$SO(3,1) = \{\Lambda \in O(3,1) \,|\, \det(\Lambda) = 1\}.$$

For example, a Lorentz transformation $\Lambda = \begin{pmatrix} 1 & 0 \\ 0 & R \end{pmatrix}$ is a proper Lorentz transformation if and only if $R \in SO(3)$ is a special orthogonal transformation. Lorentz boosts are always proper Lorentz transformations.

5.1.2 Transformation of Tensor Fields

Any tensor fields transform naturally under Lorentz transformations. For example, let

$$\Lambda : \mathbb{R}^{3,1} \longrightarrow \mathbb{R}^{3,1}.$$

Let $\alpha = \sum_\mu \alpha_\mu(x) dx^\mu$ be a 1-form on $\mathbb{R}^{3,1}$. Then α is transformed via pull-back by Λ as

$$\alpha \longmapsto \Lambda^*(\alpha).$$

Remark 5.1.5. Strictly speaking, our convention of the transformation $\alpha \to \Lambda^*(\alpha)$ defines a right action of Lorentz transformation on differential forms. Indeed, we have

$$(\Lambda_1)^*((\Lambda_2)^*(\alpha)) = (\Lambda_2 \circ \Lambda_1)^*(\alpha).$$

We could as well define $\alpha \to (\Lambda^{-1})^*(\alpha)$ for a left action. However, since we will not discuss composition of Lorentz transformations, it does not really matter. We will keep the above convention, which is convenient for differential forms, to simplify the presentation. It should be straight-forward to relate different conventions.

In coordinates, if

$$\Lambda : \mathbb{R}^{3,1} \longrightarrow \mathbb{R}^{3,1}$$
$$x^\mu \longmapsto \widetilde{x}^\mu = \sum_\nu \Lambda^\mu{}_\nu x^\nu$$

then by definition of pull-back

$$\Lambda^*(\alpha) = \sum_\mu \alpha_\mu(\widetilde{x}) d\widetilde{x}^\mu = \sum_{\mu,\nu} \alpha_\mu(\widetilde{x}) \Lambda^\mu{}_\nu dx^\nu = \sum_{\mu,\nu} \alpha_\mu(\Lambda x) \Lambda^\mu{}_\nu dx^\nu.$$

Let us denote the transformed form by $\tilde{\alpha} = \Lambda^*(\alpha)$

$$\alpha = \sum_\mu \alpha_\mu(x) dx^\mu \longmapsto \widetilde{\alpha} = \sum_\mu \widetilde{\alpha}_\mu(x) dx^\mu.$$

Expanding the relation $\tilde{\alpha} = \Lambda^*(\alpha)$ which says

$$\sum_{\mu,\nu} \alpha_\mu(\widetilde{x})\Lambda^\mu{}_\nu dx^\nu = \sum_\mu \widetilde{\alpha}_\mu(x)dx^\mu,$$

we find the transformation rule in components

$$\widetilde{\alpha}_\mu(x) = \sum_\nu \alpha_\nu(\widetilde{x})\Lambda^\nu{}_\mu = \sum_\nu \alpha_\nu(\Lambda x)\Lambda^\nu{}_\mu.$$

For simplicity, we will use Einstein summation convention, where repeated upper and lower indices refer to summation. Then under the Lorentz transform,

$$x^\mu \longmapsto \widetilde{x}^\mu = \Lambda^\mu{}_\nu x^\nu.$$

A 1-form $\alpha = \alpha_\mu(x)dx^\mu$ is transformed to

$$\alpha \longmapsto \widetilde{\alpha} = \widetilde{\alpha}_\mu(x)dx^\mu,$$

where

$$\widetilde{\alpha}_\mu(x) = \alpha_\nu(\widetilde{x})\Lambda^\nu{}_\mu.$$

Similarly, for a vector field $V = V^\mu(x)\frac{\partial}{\partial x^\mu}$, we define its transformation under Λ by

$$V \longmapsto V^\mu(\widetilde{x})\frac{\partial}{\partial \widetilde{x}^\mu} = V^\mu(\widetilde{x})\frac{\partial x^\nu}{\partial \widetilde{x}^\mu}\frac{\partial}{\partial x^\nu} = V^\mu(\widetilde{x})\left(\Lambda^{-1}\right)^\nu{}_\mu \frac{\partial}{\partial x^\nu}.$$

Let us denote transformed vector field of V by $\tilde{V}$

$$V = V^\mu(x)\frac{\partial}{\partial x^\mu} \longmapsto \widetilde{V} = \widetilde{V}^\mu(x)\frac{\partial}{\partial x^\mu}.$$

Then comparing the above two expressions, we find the transformation rule as

$$\widetilde{V}^\mu(x) = V^\nu(\widetilde{x})\left(\Lambda^{-1}\right)^\mu{}_\nu = V^\nu(\Lambda x)\left(\Lambda^{-1}\right)^\mu{}_\nu.$$

Remark 5.1.6. In geometric term, this transformation of a vector field is the push-forward

$$V \longmapsto \widetilde{V} = (\Lambda^{-1})_* V.$$

This can be further simplified using the Minkowski metric $\eta_{\mu\nu}$. We can use η to turn a form into a vector field, and vice versa. This is the standard rule to raise or lower tensor indices via η. For example, we can construct

$$V_\mu = \eta_{\mu\nu}V^\nu$$

which turns a vector field component into a 1-form component. Similarly,

$$\alpha^\mu = \eta^{\mu\nu}\alpha_\nu$$

turns a 1-form component into a vector field component. Here

$$\eta^{\mu\nu} = \left(\eta^{-1}\right)^{\mu\nu}$$

is the inverse matrix of η.

$$\eta = \begin{pmatrix} 1 & 0 & 0 & 0 \\ 0 & -1 & 0 & 0 \\ 0 & 0 & -1 & 0 \\ 0 & 0 & 0 & -1 \end{pmatrix}, \qquad \eta^{-1} = \begin{pmatrix} 1 & 0 & 0 & 0 \\ 0 & -1 & 0 & 0 \\ 0 & 0 & -1 & 0 \\ 0 & 0 & 0 & -1 \end{pmatrix}.$$

In general, for an arbitrary tensor field T with components $T^{\mu_1\cdots\mu_k}{}_{\nu_1\cdots\nu_m}$, we can raise and lower their indices by

$$T_{\mu_1\cdots\mu_k}{}^{\nu_1\cdots\nu_m} = \eta_{\mu_1\mu_1'}\cdots\eta_{\mu_k\mu_k'}\eta^{\nu_1\nu_1'}\cdots\eta^{\nu_m\nu_m'}T^{\mu_1'\cdots\mu_k'}{}_{\nu_1'\cdots\nu_m'}.$$

We can also raise or lower part of indices. The formula is similar.

With this notion at hand, we look back at the Lorentz transformation $\Lambda = (\Lambda^\mu{}_\nu)$

$$\Lambda^T\eta\Lambda = \eta, \qquad \text{or} \qquad \eta_{\mu\nu}\Lambda^\mu{}_\rho\Lambda^\nu{}_\sigma = \eta_{\rho\sigma}.$$

This is equivalently described as

$$\Lambda^\mu{}_\rho\Lambda_\mu{}^\sigma = \delta_\rho{}^\sigma.$$

In other words, the inverse matrix Λ^{-1} is precisely

$$\Lambda_\nu{}^\mu = \left(\Lambda^{-1}\right)^\mu{}_\nu.$$

Warning: Be careful about the position of indices. $(\Lambda_\nu{}^\mu)$ is the inverse matrix of $(\Lambda^\mu{}_\nu)$.

As a result, we can write the Lorentz transformation of a vector field by

$$V^\mu(x) \longmapsto \widetilde{V}^\mu(x) = V^\nu(\widetilde{x})\Lambda_\nu{}^\mu.$$

For a general tensor field

$$T = T^{\mu_1\cdots\mu_k}{}_{\nu_1\cdots\nu_m}(x)\frac{\partial}{\partial x^{\mu_1}}\otimes\cdots\otimes\frac{\partial}{\partial x^{\mu_k}}\otimes dx^{\nu_1}\otimes\cdots\otimes dx^{\nu_m},$$

it transforms under a Lorentz transformation $\Lambda : x \mapsto \widetilde{x}$ by

$$T \longmapsto \widetilde{T}$$

which is described in components via

$$\widetilde{T}^{\mu_1\cdots\mu_k}{}_{\nu_1\cdots\nu_m}(x) = T^{\mu_1'\cdots\mu_k'}{}_{\nu_1'\cdots\nu_m'}(\widetilde{x})\Lambda_{\mu_1'}{}^{\mu_1}\cdots\Lambda_{\mu_k'}{}^{\mu_k}\Lambda^{\nu_1'}{}_{\nu_1}\cdots\Lambda^{\nu_m'}{}_{\nu_m}.$$

This is the transformation rule for general tensor fields. Note that this is consistent if we raise or lower some indices of $\widetilde{T}$ (Exercise: Check this).

5.1.3 Invariance of Inner Contraction

We can change the type of a tensor field by contracting their indices. For example, consider a tensor field

$$T = T^{\mu}{}_{\nu}\frac{\partial}{\partial x^{\mu}} \otimes dx^{\nu}.$$

Then we can get a scalar (function) by

$$f(x) = T^{\mu}{}_{\mu}(x).$$

This is compatible with Lorentz transformation. In fact, under $\Lambda : x \mapsto \widetilde{x}$, the tensor field T is transformed to $\widetilde{T}$ where

$$\widetilde{T}^{\mu}{}_{\nu}(x) = T^{\mu'}{}_{\nu'}(\widetilde{x})\Lambda_{\mu'}{}^{\mu}\Lambda^{\nu'}{}_{\nu}$$

and the scalar function f is transformed to $\tilde{f}$ where

$$\widetilde{f}(x) = f(\widetilde{x}).$$

Since

$$\widetilde{T}^{\mu}{}_{\mu}(x) = T^{\mu'}{}_{\nu'}(\widetilde{x})\Lambda_{\mu'}{}^{\mu}\Lambda^{\nu'}{}_{\mu} = T^{\mu'}{}_{\nu'}(\widetilde{x})\delta_{\mu'}{}^{\nu'} = T^{\mu}{}_{\mu}(\widetilde{x}),$$

we see that $\widetilde{f}$ is again given by

$$\widetilde{f} = \widetilde{T}^{\mu}{}_{\mu}.$$

This consistency follows essentially from the invariance of Minkowski metric under Lorentz transformations.

5.2 Lorentz Invariance of Maxwell's Equations

Now we study the Lorentz transformation of electromagnetic fields and explore the invariance property of Maxwell's equations.

5.2.1 Transformation of Electromagnetic Fields

We first describe how electromagnetic fields change under Lorentz transformations. Let

$$\Lambda : \mathbb{R}^{3,1} \longrightarrow \mathbb{R}^{3,1}$$

be a Lorentz transformation which in coordinates is

$$\Lambda : x^{\mu} \longmapsto \widetilde{x}^{\mu} = \Lambda^{\mu}{}_{\nu}x^{\nu}.$$

The electric and magnetic fields are organized into a 2-form F

$$F = (\mathbf{E}_x dx + \mathbf{E}_y dy + \mathbf{E}_z dz) \wedge dt + \mathbf{B}_x dy \wedge dz + \mathbf{B}_y dz \wedge dx + \mathbf{B}_z dx \wedge dy.$$

Let us write F into tensor notation as

$$F = \frac{1}{2}\sum_{\mu,\nu} F_{\mu\nu} dx^\mu \wedge dx^\nu, \qquad F_{\mu\nu} = -F_{\nu\mu}.$$

The tensor components $F_{\mu\nu}$ are expressed in matrix form

$$F_{\mu\nu} = \begin{pmatrix} 0 & -\mathbf{E}_x/c & -\mathbf{E}_y/c & -\mathbf{E}_z/c \\ \mathbf{E}_x/c & 0 & \mathbf{B}_z & -\mathbf{B}_y \\ \mathbf{E}_y/c & -\mathbf{B}_z & 0 & \mathbf{B}_x \\ \mathbf{E}_z/c & \mathbf{B}_y & -\mathbf{B}_x & 0 \end{pmatrix} \begin{matrix} \leftarrow \mu = 0 \\ 1 \\ 2 \\ 3 \end{matrix}$$

$$\begin{matrix} \uparrow & & & \\ \nu = 0 & 1 & 2 & 3 \end{matrix}$$

Under the Lorentz transformation Λ, the tensor field $F_{\mu\nu}$ is transformed to

$$\widetilde{F}_{\mu\nu}(x) = F_{\rho\sigma}(\widetilde{x})\Lambda^\rho{}_\mu \Lambda^\sigma{}_\nu = F_{\rho\sigma}(\Lambda x)\Lambda^\rho{}_\mu \Lambda^\sigma{}_\nu.$$

This allows us to read off the transformation of electric and magnetic fields via components.

Example 5.2.1 (Lorentz boost). Consider the Lorentz boost Λ in the x-direction

$$\Lambda = \begin{pmatrix} \gamma & -\gamma v/c & 0 & 0 \\ -\gamma v/c & \gamma & 0 & 0 \\ 0 & 0 & 1 & 0 \\ 0 & 0 & 0 & 1 \end{pmatrix}$$

where $\gamma = \frac{1}{\sqrt{1-v^2/c^2}}$. Then

$$\widetilde{F}_{\mu\nu} = \begin{pmatrix} \gamma & -\gamma v/c & 0 & 0 \\ -\gamma v/c & \gamma & 0 & 0 \\ 0 & 0 & 1 & 0 \\ 0 & 0 & 0 & 1 \end{pmatrix} \begin{pmatrix} 0 & -\mathbf{E}_x/c & -\mathbf{E}_y/c & -\mathbf{E}_z/c \\ \mathbf{E}_x/c & 0 & \mathbf{B}_z & -\mathbf{B}_y \\ \mathbf{E}_y/c & -\mathbf{B}_z & 0 & \mathbf{B}_x \\ \mathbf{E}_z/c & \mathbf{B}_y & -\mathbf{B}_x & 0 \end{pmatrix}$$

$$\times \begin{pmatrix} \gamma & -\gamma v/c & 0 & 0 \\ -\gamma v/c & \gamma & 0 & 0 \\ 0 & 0 & 1 & 0 \\ 0 & 0 & 0 & 1 \end{pmatrix}$$

$$= \begin{pmatrix} 0 & -\mathbf{E}_x/c & -\gamma\mathbf{E}_y/c - \gamma\mathbf{B}_z v/c & -\gamma\mathbf{E}_z/c + \gamma\mathbf{B}_y v/c \\ \mathbf{E}_x/c & 0 & \gamma\mathbf{E}_y v/c^2 + \gamma\mathbf{B}_z & \gamma\mathbf{E}_z v/c^2 - \gamma\mathbf{B}_y \\ \gamma\mathbf{E}_y/c + \gamma\mathbf{B}_z v/c & -\gamma\mathbf{E}_y v/c^2 - \gamma\mathbf{B}_z & 0 & \mathbf{B}_x \\ \gamma\mathbf{E}_z/c - \gamma\mathbf{B}_y v/c & -\gamma\mathbf{E}_z v/c^2 + \gamma\mathbf{B}_y & -\mathbf{B}_x & 0 \end{pmatrix}.$$

It reads in components as

$$\begin{cases} \widetilde{\mathbf{E}}_x = \mathbf{E}_x \\ \widetilde{\mathbf{E}}_y = \gamma\left(\mathbf{E}_y + v\mathbf{B}_z\right) \\ \widetilde{\mathbf{E}}_z = \gamma\left(\mathbf{E}_z - v\mathbf{B}_y\right) \\ \widetilde{\mathbf{B}}_x = \mathbf{B}_x \\ \widetilde{\mathbf{B}}_y = \gamma\left(\mathbf{B}_y - v\mathbf{E}_z/c^2\right) \\ \widetilde{\mathbf{B}}_z = \gamma\left(\mathbf{B}_z + v\mathbf{E}_y/c^2\right) \end{cases}$$

where fields on the left hand side are evaluated at point x and fields on the right hand side are evaluated at the point $\widetilde{x} = \Lambda x$.

One important observation is that Lorentz transformation will mix electric and magnetic fields in general.

5.2.2 Transformation of Charge-Current Density

The charge and current densities are organized into a 1-form on $\mathbb{R}^{3,1}$

$$J = \rho/\varepsilon_0 dt - \mu_0(\mathbf{j}_x dx + \mathbf{j}_y dy + \mathbf{j}_z dz).$$

Under the Lorentz transformation $\Lambda : \mathbb{R}^{3,1} \to \mathbb{R}^{3,1}$, it transforms $J = J_\mu dx^\mu$ to $\widetilde{J} = \widetilde{J}_\mu dx^\mu$ by

$$\widetilde{J} = \Lambda^*(J)$$

or in components,

$$\widetilde{J}_\mu(x) = J_\rho(\widetilde{x})\Lambda^\rho{}_\mu = J_\rho(\Lambda x)\Lambda^\rho{}_\mu.$$

Example 5.2.2 (Lorentz boost). Under the Lorentz boost

$$\Lambda = \begin{pmatrix} \gamma & -\gamma v/c & 0 & 0 \\ -\gamma v/c & \gamma & 0 & 0 \\ 0 & 0 & 1 & 0 \\ 0 & 0 & 0 & 1 \end{pmatrix},$$

we find

$$\begin{cases} \widetilde{\rho} = \gamma\left(\rho + \mathbf{j}_x v/c^2\right) \\ \widetilde{\mathbf{j}}_x = \gamma\left(\mathbf{j}_x + \rho v\right) \\ \widetilde{\mathbf{j}}_y = \mathbf{j}_y \\ \widetilde{\mathbf{j}}_z = \mathbf{j}_z \end{cases}$$

Again, fields on the left hand side are evaluated at point x and fields on the right hand side are evaluated at the point $\widetilde{x} = \Lambda x$.

5.2.3 Transformation of Maxwell's Equations

The Maxwell's equations are

$$\begin{cases} dF = 0 \\ d(\star F) = \star J \end{cases}$$

To understand the Lorentz transformation of Maxwell's equations, we need to know how the Hodge star $\star$ intertwines with the Lorentz transformation. Note that we use the notation $\star$ for Hodge star in this subsection, in order to distinguish with the $*$-symbol in pull-backs.

Proposition 5.2.3. *Let $\Lambda : \mathbb{R}^{3,1} \to \mathbb{R}^{3,1}$ be a proper Lorentz transformation. Then for any differential form $\alpha \in \Omega^\bullet(\mathbb{R}^{3,1})$, we have*

$$\Lambda^*(\star\alpha) = \star\Lambda^*(\alpha).$$

In other words, the Hodge star $\star$ commutes with the pull-back via Λ.

Proof: It is enough to check on the basis

$$1, \qquad dx^\mu, \qquad dx^\mu \wedge dx^\nu, \qquad dx^\mu \wedge dx^\nu \wedge dx^\rho, \qquad dx^0 \wedge dx^1 \wedge dx^2 \wedge dx^3.$$

We check one case for $\alpha = 1$. The others are similar.

$$\Lambda^*(\star 1) = \Lambda^*(dx^0 \wedge dx^1 \wedge dx^2 \wedge dx^3) = \det(\Lambda)(dx^0 \wedge dx^1 \wedge dx^2 \wedge dx^3)$$

and

$$\star\Lambda^*(1) = \star 1 = dx^0 \wedge dx^1 \wedge dx^2 \wedge dx^3.$$

For a proper Lorentz transformation, $\det(\Lambda) = 1$. Thus the above two terms are equal. □

Remark 5.2.4. For a general Lorentz transformation Λ, we have

$$\Lambda^*(\star\alpha) = (\det\Lambda) \star \Lambda^*(\alpha).$$

The proof is similar.

We will now focus on proper Lorentz transformations $\Lambda \in SO(3,1)$, which preserves the orientation on $\mathbb{R}^{3,1}$. Under the Lorentz transformation by Λ, we have

$$\begin{cases} \widetilde{F} = \Lambda^*(F) \\ \widetilde{J} = \Lambda^*(J) \end{cases}$$

Since the pull-back Λ^* commutes with d, and also commutes with the Hodge star $\star$ as in Proposition 5.2.3, we find

$$\begin{cases} dF = 0 \\ d(\star F) = \star J \end{cases} \quad \xRightarrow[\text{via } \Lambda]{\text{pull-back}} \quad \begin{cases} d\widetilde{F} = 0 \\ d(\star\widetilde{F}) = \star\widetilde{J} \end{cases}$$

Thus the transformed electromagnetic field and the transformed charge-current density again satisfy the Maxwell's equations! In other words, Lorentz transformations will transform solutions of Maxwell's equations to solutions of Maxwell's equations.

Another illustrating way to see this is to consider the Maxwell action

$$S_M[A, J] = \frac{1}{\mu_0}\int_{\mathbb{R}^{3,1}} \left(\frac{1}{2}F \wedge \star F + \star J \wedge A\right).$$

Under the Lorentz transformation Λ, the gauge 1-form $A = A_\mu dx^\mu$ transforms as

$$A \longmapsto \widetilde{A}_\mu(x)dx^\mu = \Lambda^*(A)$$

or in components $\widetilde{A}_\mu(x) = A_\rho(\widetilde{x})\Lambda^\rho{}_\mu$. Then

$$\begin{aligned} S_M[\widetilde{A}, \widetilde{J}] &= \frac{1}{\mu_0}\int_{\mathbb{R}^{3,1}} \left(\frac{1}{2}\widetilde{F} \wedge \star\widetilde{F} + \star\widetilde{J} \wedge \widetilde{A}\right) \\ &= \frac{1}{\mu_0}\int_{\mathbb{R}^{3,1}} \Lambda^* \left(\frac{1}{2}F \wedge \star F + \star J \wedge A\right) \\ &= \frac{1}{\mu_0}\int_{\mathbb{R}^{3,1}} \left(\frac{1}{2}F \wedge \star F + \star J \wedge A\right) = S_M[A, J]. \end{aligned}$$

Here in the second line, we have used the invariance of integral under orientation-preserving diffeomorphisms. It follows that A is a critical point of $S_M[A, J]$ if and only if $\widetilde{A}$ is a critical point of $S_M[\widetilde{A}, \widetilde{J}]$, i.e. Lorentz transformations send solutions of Maxwell's equations to solutions.

Example 5.2.5. Consider the electromagnetic field of a static point particle of charge q at the origin

$$\vec{\mathbf{E}} = \frac{q}{4\pi\varepsilon_0}\frac{\vec{r}}{r^3}, \qquad \vec{\mathbf{B}} = 0.$$

Under the Lorentz boost $\Lambda : \mathbb{R}^{3,1} \to \mathbb{R}^{3,1}$

$$\Lambda = \begin{pmatrix} \gamma & -\gamma v/c & 0 & 0 \\ -\gamma v/c & \gamma & 0 & 0 \\ 0 & 0 & 1 & 0 \\ 0 & 0 & 0 & 1 \end{pmatrix}$$

with

$$\begin{cases} c\widetilde{t} = \dfrac{ct - xv/c}{\sqrt{1 - v^2/c^2}} \\ \widetilde{x} = \dfrac{x - vt}{\sqrt{1 - v^2/c^2}} \\ \widetilde{y} = y \\ \widetilde{z} = z \end{cases}$$

We find the transformed fields by

$$\begin{cases} \widetilde{\mathbf{E}}_x = \dfrac{q}{4\pi\varepsilon_0}\dfrac{\widetilde{x}}{\widetilde{r}^3} = \dfrac{q}{4\pi\varepsilon_0}\dfrac{\gamma(x - vt)}{(\gamma^2(x - vt)^2 + y^2 + z^2)^{3/2}} \\ \widetilde{\mathbf{E}}_y = \dfrac{q}{4\pi\varepsilon_0}\dfrac{\gamma\widetilde{y}}{\widetilde{r}^3} = \dfrac{q}{4\pi\varepsilon_0}\dfrac{\gamma y}{(\gamma^2(x - vt)^2 + y^2 + z^2)^{3/2}} \\ \widetilde{\mathbf{E}}_z = \dfrac{q}{4\pi\varepsilon_0}\dfrac{\gamma\widetilde{z}}{\widetilde{r}^3} = \dfrac{q}{4\pi\varepsilon_0}\dfrac{\gamma z}{(\gamma^2(x - vt)^2 + y^2 + z^2)^{3/2}} \\ \widetilde{\mathbf{B}}_x = 0 \\ \widetilde{\mathbf{B}}_y = -\dfrac{q}{4\pi\varepsilon_0}\dfrac{\gamma v}{c^2}\dfrac{\widetilde{z}}{\widetilde{r}^3} = \dfrac{q}{4\pi\varepsilon_0}\dfrac{\gamma vz/c^2}{(\gamma^2(x - vt)^2 + y^2 + z^2)^{3/2}} \\ \widetilde{\mathbf{B}}_z = \dfrac{q}{4\pi\varepsilon_0}\dfrac{\gamma v}{c^2}\dfrac{\widetilde{y}}{\widetilde{r}^3} = \dfrac{q}{4\pi\varepsilon_0}\dfrac{\gamma vy/c^2}{(\gamma^2(x - vt)^2 + y^2 + z^2)^{3/2}} \end{cases}$$

On the other hand, the charge-current densities

$$\begin{cases} \rho = q\delta(\vec{r}) \\ \vec{\mathbf{j}} = 0 \end{cases}$$

is transformed to

$$\begin{cases} \widetilde{\rho} = \gamma q\delta\left(\vec{\widetilde{r}}\right) = \gamma q\delta(\gamma(x-vt))\delta(y)\delta(z) = q\delta(x-vt)\delta(y)\delta(z) \\ \widetilde{\mathbf{j}}_x = qv\delta(x-vt)\delta(y)\delta(z) \\ \widetilde{\mathbf{j}}_y = \widetilde{\mathbf{j}}_z = 0 \end{cases}$$

This describes a point charge moving in the x-direction with constant velocity v. We have recovered the Liénard-Wiechert electric and magnetic fields for a point charge of constant velocity described in Example 5.2.5.

5.3 Relativistic Lorentz Force Law

We discuss relativistic version of Lorentz force law.

5.3.1 Non-relativistic Charged Particle

A particle of charge q moving with velocity $\vec{v}$ in the background of electromagnetic fields will experience a force via the Lorentz force law

$$\vec{F} = q\left(\vec{\mathbf{E}} + \vec{v} \times \vec{\mathbf{B}}\right).$$

Assume the particle has mass m, and the trajectory is described by $\vec{r}(t) = (x(t), y(t), z(t))$. Then the equation of motion is described via Newton's law by

$$m\ddot{\vec{r}} = q\left(\vec{\mathbf{E}} + \dot{\vec{r}} \times \vec{\mathbf{B}}\right).$$

This equation can be described by a Lagrangian

$$\mathscr{L} = \frac{1}{2}m\dot{\vec{r}}^2 - q\phi(\vec{r}, t) + q\vec{\mathbf{A}}(\vec{r}, t) \cdot \dot{\vec{r}}.$$

Here ϕ is the scalar potential and $\vec{\mathbf{A}}$ is the vector potential. Then the Euler-Lagrange equation of the action

$$S[\vec{r}] = \int dt\, \left(\frac{1}{2}m\dot{\vec{r}}^2 - q\phi + q\vec{\mathbf{A}} \cdot \dot{\vec{r}}\right)$$

gives precisely (for example, see [10, Example 1.2.3] in this series) the above Lorentz force law

$$\delta S = 0 \qquad \Longrightarrow \qquad m\ddot{\vec{r}} = q\left(\vec{\mathbf{E}} + \dot{\vec{r}} \times \vec{\mathbf{B}}\right).$$

5.3.2 Relativistic Charged Particle

Let us now work with a relativistic particle of mass m and charge q, whose trajectory in the spacetime is described by

$$\gamma = (x^0(\sigma), x^1(\sigma), x^2(\sigma), x^3(\sigma)).$$

Here σ is a parameter of the curve γ in $\mathbb{R}^{3,1}$.

The relativistic action for the charged particle in the electromagnetic background is

$$S[\gamma] = -mc \int_\gamma \sqrt{\eta_{\mu\nu} dx^\mu dx^\nu} + q \int_\gamma A.$$

Let us explain these two terms.

① In terms of a parameter σ on γ, the first term is expressed explicitly as

$$-mc \int d\sigma \sqrt{\eta_{\mu\nu} \frac{dx^\mu}{d\sigma} \frac{dx^\nu}{d\sigma}}.$$

It is clear from the expression that this term does not depend on the choice of parameter σ. For another parameter $\widetilde{\sigma}$ of the curve γ,

$$\begin{aligned} \int d\sigma \sqrt{\eta_{\mu\nu} \frac{dx^\mu}{d\sigma} \frac{dx^\nu}{d\sigma}} &= \int d\sigma \sqrt{\eta_{\mu\nu} \frac{dx^\mu}{d\widetilde{\sigma}} \frac{dx^\nu}{d\widetilde{\sigma}} \left(\frac{d\widetilde{\sigma}}{d\sigma}\right)^2} = \int d\sigma \left|\frac{d\widetilde{\sigma}}{d\sigma}\right| \sqrt{\eta_{\mu\nu} \frac{dx^\mu}{d\widetilde{\sigma}} \frac{dx^\nu}{d\widetilde{\sigma}}} \\ &= \int d\widetilde{\sigma} \sqrt{\eta_{\mu\nu} \frac{dx^\mu}{d\widetilde{\sigma}} \frac{dx^\nu}{d\widetilde{\sigma}}}. \end{aligned}$$

This also explains why we write the action as

$$-mc \int_\gamma \sqrt{\eta_{\mu\nu} dx^\mu dx^\nu}$$

without using any explicit parameter.

If we use the time t as a parameter, then the curve is parameterized as

$$\begin{aligned} \gamma(t) &= (x^0(t), x^1(t), x^2(t), x^3(t)) \\ &= (ct, x(t), y(t), z(t)) \\ &= (ct, \vec{r}(t)) \end{aligned}$$

where $\vec{r}(t) = (x(t), y(t), z(t))$. Then the action becomes

$$\begin{aligned} -mc \int dt \sqrt{c^2 - \dot{\vec{r}}^2} &= -mc^2 \int dt \sqrt{1 - \dot{\vec{r}}^2/c^2} \\ &= -mc^2 \int dt \left(1 - \frac{1}{2}\frac{\dot{\vec{r}}^2}{c^2} + \cdots\right) \\ &= \text{const} + \frac{1}{2} m \int dt\, \dot{\vec{r}}^2 + \cdots \end{aligned}$$

Here terms in $\cdots$ involve higher orders in $\dot{\vec{r}}/c$, which become zero in the non-relativistic limit. It follows that the term

$$-mc\int_\gamma \sqrt{\eta_{\mu\nu}dx^\mu dx^\nu}$$

becomes the standard kinetic term $\frac{1}{2}m\int dt\,\dot{\vec{\gamma}}^2$ in the non-relativistic limit, up to a constant which is irrelevant for equation of motion. Note that $\eta_{\mu\nu}dx^\mu dx^\nu$ is always nonnegative since particles can not travel faster than the speed of light.

② In the second term,

$$A = -\phi dt + \mathbf{A}_x dx + \mathbf{A}_y dy + \mathbf{A}_z dz$$

is the gauge 1-form, which can be integrated along a curve γ in $\mathbb{R}^{3,1}$. This is $q\int_\gamma A$.

If we parametrize the curve γ via time t along the curve

$$\gamma(t) = (ct, \vec{r}(t)),$$

then

$$q\int_\gamma A = q\int dt\left(-\phi + \mathbf{A}_x\frac{dx}{dt} + \mathbf{A}_y\frac{dy}{dt} + \mathbf{A}_z\frac{dz}{dt}\right) = \int dt\left(-q\phi + q\vec{\mathbf{A}}\cdot\dot{\vec{r}}\right).$$

This coincides with the potential term in the case of non-relativistic charged particle.

The above shows that

$$S[\gamma] = -mc\int_\gamma \sqrt{\eta_{\mu\nu}dx^\mu dx^\nu} + q\int_\gamma A$$

is indeed a natural relativistic generation of action functional for charged particle.

Let us now work out the equation of motion for $S[\gamma]$, which would describe the relativistic Lorentz force law. Let us parametrize the curve as

$$\gamma(\sigma) = \{x^\mu(\sigma)\}$$

and the action becomes

$$S = -mc\int d\sigma\sqrt{\eta_{\mu\nu}\frac{dx^\mu}{d\sigma}\frac{dx^\nu}{d\sigma}} + q\int d\sigma\, A_\mu\frac{dx^\mu}{d\sigma}.$$

Consider an arbitrary variation

$$\gamma \longrightarrow \gamma + \delta\gamma.$$

The variation of the action is

$$\begin{aligned}
\delta S &= -mc\int d\sigma\frac{\eta_{\mu\nu}\frac{d\delta x^\mu}{d\sigma}\frac{dx^\nu}{d\sigma}}{\sqrt{\eta_{\mu\nu}\frac{dx^\mu}{d\sigma}\frac{dx^\nu}{d\sigma}}} + q\int d\sigma\, A_\mu\frac{d\delta x^\mu}{d\sigma} + q\int d\sigma\,(\partial_\mu A_\nu)\delta x^\mu\frac{dx^\nu}{d\sigma} \\
&= mc\int d\sigma\,\delta x^\mu\frac{d}{d\sigma}\left(\frac{\eta_{\mu\nu}\frac{dx^\nu}{d\sigma}}{\sqrt{\eta_{\mu\nu}\frac{dx^\mu}{d\sigma}\frac{dx^\nu}{d\sigma}}}\right) + q\int d\sigma\,\delta x^\mu\left(\partial_\mu A_\nu - \partial_\nu A_\mu\right)\frac{dx^\nu}{d\sigma} \\
&= \int d\sigma\,\delta x^\mu\left\{\frac{d}{d\sigma}\left(\frac{mc\eta_{\mu\nu}\frac{dx^\nu}{d\sigma}}{\sqrt{\eta_{\mu\nu}\frac{dx^\mu}{d\sigma}\frac{dx^\nu}{d\sigma}}}\right) + qF_{\mu\nu}\frac{dx^\nu}{d\sigma}\right\}.
\end{aligned}$$

The equation of motion is obtained by asking $\delta S = 0$ for an arbitrary variation δx^μ. This leads to

$$\frac{d}{d\sigma} P_\mu = -qF_{\mu\nu} \frac{dx^\nu}{d\sigma}$$

where $P_\mu = \eta_{\mu\nu} P^\nu$ and

$$P^\nu = mc \frac{\frac{dx^\nu}{d\sigma}}{\sqrt{\eta_{\rho\nu} \frac{dx^\rho}{d\sigma} \frac{dx^\nu}{d\sigma}}}.$$

Note that P^μ is invariant under the change of the parametrization σ, by the same reason as above. It is precisely the relativistic momentum. In fact, let us choose the time t as the parameter, then

$$P^\mu = mc \frac{\dot{x}^\mu}{\sqrt{c^2 - \dot{\vec{r}}^2}} = \left(\frac{mc}{\sqrt{1 - \dot{\vec{r}}^2/c^2}}, \frac{m\dot{\vec{r}}}{\sqrt{1 - \dot{\vec{r}}^2/c^2}} \right)$$

where $\vec{r} = (x(t), y(t), z(t))$. We see that

$$P^\mu P_\mu = (P^0)^2 - (P^1)^2 - (P^2)^2 - (P^3)^2 = m^2c^2$$

which is the familiar result on relativistic 4-momentum. cP^0 gives the relativistic energy

$$\frac{mc^2}{\sqrt{1 - \dot{\vec{r}}^2/c^2}} = mc^2 + \frac{1}{2} m\dot{\vec{r}}^2 + \cdots$$

and $\frac{1}{2} m\dot{\vec{r}}^2$ is the non-relativistic kinetic energy. The non-relativistic limit of $\frac{m\dot{\vec{r}}}{\sqrt{1-\dot{\vec{r}}^2/c^2}}$ gives $m\dot{\vec{r}}$, which is the standard momentum vector.

It is illustrating to unpack the extra equation coming from $\mu = 0$:

$$\frac{dP_0}{dt} = -qF_{0\nu} \frac{dx^\nu}{dt} = q\vec{\mathbf{E}} \cdot \dot{\vec{r}}/c,$$

i.e.

$$\frac{d(cP_0)}{dt} = q\vec{\mathbf{E}} \cdot \dot{\vec{r}}.$$

Since cP_0 is the energy, this equation simply tells the change of kinetic energy when work is done by an electric field.

Bibliography

[1] Berger, Mitchell A., and George B. Field. *The topological properties of magnetic helicity.* Journal of Fluid Mechanics 147 (1984): 133–148.

[2] Buchwald, Jed Z. *Electrodynamics from Thomson and Maxwell to Hertz.* The Oxford Handbook of the History of Physics. Oxford University Press, 2013.

[3] Butcher, Ginger. *Tour of the Electromagnetic Spectrum.* Government Printing Office, 2016.

[4] Eguchi, Tohru, Peter B. Gilkey, and Andrew J. Hanson. *Gravitation, Gauge Theories and Differential Geometry.* Physics Reports 66.6 (1980): 213–393.

[5] Garrity, Thomas A. *Electricity and Magnetism for Mathematicians: A Guided Path from Maxwell's Equations to Yang-Mills.* Cambridge University Press, 2015.

[6] Göckeler, Meinulf, and Thomas Schücker. *Differential Geometry, Gauge Theories, and Gravity.* Cambridge University Press, 1989.

[7] Griffiths, David J., and Reed Colleger. *Introduction to Electrodynamics.* Prentice Hall Upper Saddle River, New Jersey 7458 (1999).

[8] Jackson, John David. *Classical Electrodynamics.* John Wiley & Sons, 1999.

[9] Jackson, John David, and Lev Borisovich Okun. *Historical roots of gauge invariance.* Reviews of Modern Physics 73.3 (2001): 663.

[10] Li, Si. *Classical Mechanics and Geometry.* International Press of Boston, 2023.

[11] Maxwell, James Clerk. *A Treatise on Electricity and Magnetism.* Oxford: Clarendon Press, 1873.

[12] Nakahara, Mikio. *Geometry, Topology and Physics.* CRC Press, 2003.

[13] Nayfeh, Munir H., and Morton K. Brussel. *Electricity and Magnetism.* Courier Dover Publications, 2015.

[14] O'Raifeartaigh, Lochlainn. *The Dawning of Gauge Theory.* Vol. 106. Princeton University Press, 1997.

[15] Schwinger, Julian, Lester L. Deraad Jr., Kimball A. Milton, Wu-yang Tsai. *Classical Electrodynamics.* Westview Press, 1998.

[16] Tong, David. *Electromagnetism.* University of Cambridge Mathematical Tripos. 2015.

[17] Weyl, Hermann. *Space–Time–Matter.* Dutton, 1922.

[18] Zangwill, Andrew. *Modern Electrodynamics.* Cambridge University Press, 2013.